AGR 7514

MULTIRATE SWITCHED-CAPACITOR CIRCUITS FOR 2-D SIGNAL PROCESSING

MULTIRATE SWITCHED-CAPACITOR CIRCUITS FOR 2-D SIGNAL PROCESSING

by

Wang Ping
Instituto Superior Tecnico,
University of Lisbon

and

José E. Franca
Instituto Superior Tecnico,
University of Lisbon

KLUWER ACADEMIC PUBLISHERS
BOSTON / DORDRECHT / LONDON

A C.I.P. Catalogue record for this book is available from the Library of Congress.

ISBN 0-7923-8051-7

Published by Kluwer Academic Publishers,
P.O. Box 17, 3300 AA Dordrecht, The Netherlands.

Sold and distributed in the U.S.A. and Canada
by Kluwer Academic Publishers,
101 Philip Drive, Norwell, MA 02061, U.S.A.

In all other countries, sold and distributed
by Kluwer Academic Publishers,
P.O. Box 322, 3300 AH Dordrecht, The Netherlands.

Printed on acid-free paper

All Rights Reserved
© 1998 Kluwer Academic Publishers
No part of the material protected by this copyright notice may be reproduced or
utilized in any form or by any means, electronic or mechanical,
including photocopying, recording or by any information storage and
retrieval system, without written permission from the copyright owner.

Printed in the Netherlands

To

Wang Cheng-Da, Wu Xue-Yi

CONTENTS

List of Figures ix
List of Tables xv
Abbreviated Words xvii
List of Symbol xix
Preface xxi

1 INTRODUCTION 1
    1.1 Introductory Remarks 1
    1.2 Book Outline 3
    REFERENCES 5

2 2-D SIGNALS AND FILTERING SYSTEMS 9
    2.1 Introduction 9
    2.2 2-D Signals 10
    2.3 2-D Image Filtering Systems 15
    2.4 Hardware Implementation of 2-D Filters 23
    2.5 Application Examples of 2-D Filter Systems 29
    2.6 Summary 35
    REFERENCES 36

3 FUNDAMENTAL ASPECTS OF 2-D DECIMATION FILTERS 39
    3.1 Introduction 39
    3.2 1-D Decimation 40
    3.3 2-D Decimation 45
    3.4 Delay-Line Memory Requirements for 2-D Filters 52
    3.5 Summary 55
    REFERENCES 56

4 POLYPHASE-COEFFICIENT STRUCTURES FOR 2-D DECIMATION FILTERS — 59

- 4.1 Introduction — 59
- 4.2 Modified 1-D Decimation Polyphase Structures — 60
- 4.3 Polyphase-Coefficient Structures for 2-D Decimation Filters — 64
- 4.4 Summary — 69
- REFERENCES — 71

5 SC ARCHITECTURES FOR 2-D DECIMATION FILTERS IN POLYPHASE-COEFFICIENT FORM — 73

- 5.1 Introduction — 73
- 5.2 SC Building Blocks — 73
- 5.3 SC Architectures with 2-D Polyphase-Coefficients — 76
- 5.4 Design Example of an FIR 2-D Decimation Filter — 89
- 5.5 Design Example of an IIR 2-D Decimation Filter — 92
- 5.6 Summary — 96
- REFERENCES — 97

6 A REAL-TIME 2-D ANALOG MULTIRATE IMAGE PROCESSOR IN 1.0-μm CMOS TECHNOLOGY — 99

- 6.1 Introduction — 99
- 6.2 2-D Multirate Filter Design — 99
- 6.3 SC Horizontal Decimation Filter — 102
- 6.4 SC Vertical Filter and Associated Delay-Line Memory Blocks — 105
- 6.5 Integrated Circuit Implementation — 111
- 6.6 Experimental Characterization — 115
- 6.7 Summary — 122
- REFERENCES — 124

List of Figures

CHAPTER 2

Fig. 2.1   Illustration of (a) a 2-D image signal represented by a sequence of lines and (b) a frame representation of the image in the $[t, nT]$ space.

Fig. 2.2   (a) Frequency spectrum of a 2-D signal in 3-D form and (b) its topview representation.

Fig. 2.3   (a) 2-D sampled image signal for each line scan and (b) a frame representation of the image.

Fig. 2.4   (a) Illustration of the frequency spectrum of the 2-D signal in Fig. 2.3. (b) Topview version of (a).

Fig. 2.5   2-D analog filter with continuous-time line processing.

Fig. 2.6   Direct-form realization of a (2 x 2)nd-order analog filter. (a) Block-diagram of the overall 2-D analog filter and (b) possible realization of $Y_0(s, nT)$.

Fig. 2.7   2-D discrete-time filter with discrete-time line processing.

Fig. 2.8   Block diagram of a separable 2-D digital filter.

Fig. 2.9   Direct-form architecture for the realization of a 2-D FIR filtering system.

Fig. 2.10 Direct-form realization of a 2-D IIR filtering system with separable denominator.

Fig. 2.11 A general 2-D digital filter implementation scheme.

Fig. 2.12 A general 2-D analog filter implementation scheme.

Fig. 2.13 Outline of the chip for the (a) 2-D analysis filter, and (b) 2-D synthesis filter.

Fig. 2.14   Chip outline of (a) filter chip and (b) DL chip.

Fig. 2.15   (a) Block diagram and (b) chip outline of an analog video comb filter employing SC techniques for NTSC applications [2.39].

Fig. 2.16   (a) Block diagram and (b) chip outline of an analog 2-D highpass filter employing SC techniques [2.10].

Fig. 2.17   Typical amplitude response of a 2-D lowpass filter system for noise remove.

Fig. 2.18   (a) Ideal image in 3-D view; (b) image of (a) added by high-frequency noise; (c) resulting image after noise removal processing.

Fig. 2.19   Typical amplitude response of a 2-D highpass filtering system for edge enhancement.

Fig. 2.20   (a) Ideal image. (b) Processed image with enhanced edges.

Fig. 2.21   Block diagram of a subband codec using a separable 2-D filter bank.

Fig. 2.22   Frequency band division into 4 subbands.

Fig. 2.23   Example of band separation using 2-D subband filtering.

Fig. 2.24   Block diagram of a video compression system employing image resizing functions.

CHAPTER 3

Fig. 3.1 M-fold downsampler. (a) Symbolic representation and (b) time-domain interpretation for $M = 2$.

Fig. 3.2   The downsampling of a discrete-time signal translates the signal to lower frequencies.

Fig. 3.3   Representation of (a) the general block diagram of a decimator and (b) its efficient form of implementation to take advantage of the reduced output sampling frequency.

Fig. 3.4   1-D polyphase structure for an FIR decimation filter with M-fold sampling rate reduction.

List of Figures

Fig. 3.5   1-D polyphase structure for an IIR decimation filter with M-fold sampling rate reduction.

Fig. 3.6   2-D downsampler with decimation ratio (M_1, M_2).

Fig. 3.7   (a) Original discrete-time 2-D input image signal. (b) Decimated 2-D image, by $M_2 = 3$, in the horizontal dimension. (c) Resulting compressed frequency spectrum.

Fig. 3.8   (a) Decimation occurring in both dimensions with $M_1 = 2$, $M_2 = 3$. (b) Resulting frequency spectrum.

Fig. 3.9   (a) A general implementation of the 2-D decimation filter and (b) efficient implementation of such system.

Fig. 3.10   Polyphase implementation of a 2-D FIR decimation filter.

Fig. 3.11   Efficient polyphase implementation for a 2-D separable decimation filter.

Fig. 3.12   Symbolic representation of a DL memory block.

Fig. 3.13   (a) Traditional 1-D filtering system cascaded with a delay chain of 10 unit-delays. (b) Equivalent multirate operation. (c) Moving the downsampler to the front of the delay chain. (d) Moving the downsampler to the front of the decimation filter.

CHAPTER 4

Fig. 4.1   1-D direct-form polyphase-coefficient structure for 1-D decimation filtering.

Fig. 4.2   1-D ADB polyphase-coefficient structure for FIR decimation filtering.

Fig. 4.3   1-D ADB polyphase-coefficient structure for IIR decimation filtering.

Fig. 4.4   FIR Polyphase-coefficient structure for 2-D decimation filters with length $(N_1 \times N_2)$.

Fig. 4.5   Polyphase-coefficient structure for 2-D IIR decimation filters with length $(N_1 \times N_2)$.

CHAPTER 5

Fig. 5.1   Basic SC building blocks and switch timing. (a) Weighting branch without delay. (b) Weighting branch with one unit-delay. (c) Weighting branch with two unit-delays. (d) Weighting branch with two negative unit-delays. (e) Backward-Euler integrator with three unit-delay sampling periods. (f) Accumulation block to accumulate input samples during three phases. (g) Switch timing.

Fig. 5.2   An illustrative example of a 2-D SC filter circuit with 2nd-order filtering, $N_1 = N_2 = 2$, and decimation of $M_1 = M_2 = 3$. (a) Architecture, (b) schematic diagram, and (c) associated switch timing.

Fig. 5.3   Unfolded operation of the SC circuit of Fig. 5.2 during (a) switching phase L2, (b) switching phase L1 and (c) switching phase L0.

Fig. 5.4   Example of a 2-D SC FIR decimation filter with $N_1 = 2$, $N_2 = 3$ and $M_1 = M_2 = 3$. (a) Architecture and (b) schematic diagram.

Fig. 5.5   Example of a 2-D SC FIR decimation filter with $N_1 = 4$, $N_2 = 2$ and $M_1 = M_2 = 3$. (a) Architecture and (b) SC implementation.

Fig. 5.6   Unfolded operation of the SC circuit of Fig. 5.5. (a) Switching phase L2. (b) Switching phase L1. (c) Switching phase L0, in which the synchronized operation of charge transfer from the 2nd to the 1st active DL blocks takes place.

Fig. 5.7   An example of a 2-D SC FIR decimation filter with $N_1 = 4$, $N_2 = 3$ and $M_1 = M_2 = 3$. (a) Architecture and (b) schematic diagram.

Fig. 5.8   (a) SC network and (b) controlling switching waveforms for the realization of a 2-D FIR decimation filtering function.

Fig. 5.9   Computer simulated amplitude responses of the 2-D SC decimating circuit of Fig. 5.8. (a) Impulse sampled. (b) Sampled-and-hold effect in both dimensions.

Fig. 5.10 (a) Schematic of the 2-D SC decimator with IIR filtering. (b) Switching waveforms.

Fig. 5.11 Computer simulated amplitude responses of the 2-D SC decimation filter with IIR filtering function. (a) Impulse sampling. (b) Sampling-and-hold effect in the horizontal dimension.

CHAPTER 6

Fig. 6.1 (a) The 2-D image processor comprises an horizontal lowpass decimating filter and a vertical lowpass filter. (b) The line memory depth of the vertical filter is halved due to the horizontal 2-fold decimating filtering function.

Fig. 6.2 Circuit diagram and capacitance values (in pF) of the fully-differential SC horizontal decimation filter.

Fig. 6.3 (a) Simplified (single-ended) circuit diagram of the DL memory block and corresponding (b) Write and (c) Read modes.

Fig. 6.4 Fully-differential DL memory block. (a) Architecture and (b) storage cell. (c) Multiplexer circuit for writing. (d) Demultiplexer circuit for reading.

Fig. 6.5 DL address generation circuit.

Fig. 6.6 Circuit diagram and capacitance values (in pF) of the fully-differential SC vertical filter.

Fig. 6.7 Simplified DL memory block circuit in the Write mode with parasitic capacitance and resistance.

Fig. 6.8 (a) Fully-differential folded-cascoded OTA with (b) master current biasing circuit.

Fig. 6.9 (a) Block diagram and (b) chip floor plan of the complete 2-D multirate image processor.

Fig. 6.10 Microphotograph of the prototype 2-D filter.

Fig. 6.11 Experimental set-up for chip testing and characterization.

Fig. 6.12 Frequency amplitude responses of the implemented horizontal decimation filter.

Fig. 6.13 Measured (a) amplitude responses and (b) single-frequency response of the DL memory block.

Fig. 6.14 Experimental frequency responses of the 2-D decimation filter.

Fig. 6.15   Paulo's images. (a) Original PAL image and (b) its highpass filtered version from the prototype chip.

Fig. 6.16   The forehead images. (a) Input PAL image and (b) output highpass version from the prototype chip.

List of Tables

Table 2.1	Performance characteristics of the 2-D filter chip in [2.37].
Table 2.2	Summary of the filter chip performance [2.38].
Table 2.3	Summary of the programmable DL chip performance[2.38].
Table 2.4	Performance characteristics of the 2-D analog filter chip in [2.39].
Table 2.5	Performance characteristics of the 2-D analog filter chip in [2.10].
Table 4.1	Typical requirements of multirate and traditional (non-multirate) 2-D filtering in NTSC systems.
Table 5.1	Specifications of the 2-D FIR filter.
Table 5.2	Impulse response coefficients of the 2-D FIR prototype filter function.
Table 5.3	Specifications of the 2-D IIR decimating filter example.
Table 5.4	Coefficients of the prototype 2-D IIR filter function.
Table 5.5	Coefficients of the modified 2-D decimating filter function in the z_2 dimension.
Table 5.6	Coefficients of the modified 2-D decimating function in the z_1 dimension.
Table 6.1	Specifications of the horizontal decimating filter.
Table 6.2	Specifications of the vertical filter.
Table 6.3	Summary of measured performance characteristics of the CMOS prototype chip.
Table 6.4	Cost of implementation of the fully analog real-time image processing chip and the conversion blocks that would be needed to support a fully digital processing core.

Abbreviated Words

1-D	one-Dimensional
2-D	two-Dimensional
3-D	three-Dimensional
A/D	analog-to-digital
ADB	active-delayed-block
ARAM	analog random access memory
ASIC	application specific integrated circuit
CCD	charge coupled device
CCIR601	a standard from international radio consultative committee to produce a digital signal that is more compatible with NTSC, PAL and SECAM video formats
CIF	common intermediate format
CMOS	complementary metal oxide semiconductor
D/A	digital-to-analog
DL	delay line
DSP	digital signal processing
FIR	finite impulse response
FFT	fast Fourier transform
HDTV	high definition television

HP	highpass
IC	integrated circuit
IIR	infinite impulse response
I/P	input
LCD	liquid crystal display
LP	lowpass
LTI	linear time-invariant
MUX	multiplex
NTSC	national television system committee
O/P	output
OTA	operational transconductance amplifier
PAL	phase alternation line
QCIF	quarter common intermediate format
R/W	read or write
S/H	sample-and-hold
SC	switched-capacitor
SECAM	Sequentiel Couleur avec mémoire
VLSI	very large scale integration

List of Symbols

F_{s1}	sampling frequency in the vertical dimension
F_{s2}	sampling frequency in the horizontal dimension
g_m	transconductance
N	filter length
M	decimation ratio
OP_w	operational amplifier to write data into a DL memory
OP_r	operational amplifier to read data from a DL memory
s_1	Laplace-transform variable in the vertical dimension
s_2	Laplace-transform variable in the horizontal dimension
Vp-p	peak-to-peak voltage value
z_1	z-transform variable in the vertical dimension
z_2	z-transform variable in the horizontal dimension
z_1^{-1}	vertical unit-delay (DL)
z_2^{-1}	horizontal unit-delay

Preface

Traditional two-dimensional (2-D) switched-capacitor (SC) filters require very large delay-line (DL) memory blocks mainly because of the oversampled characteristic of the signal, both at the input and at the output. By bringing the output sampling rate closer to the Nyquist signal bandwidth it is possible to reduce considerably the size of the DL memory blocks, and thus saving silicon area for more economical integrated circuit implementation. Besides, lower output sampling rates also contribute easier signal digitization and processing and economy of signal transmission. Such operation of reducing the sampling rate from the input to the output can be achieved using appropriate 2-D decimating filters which, besides the required baseband filtering, should also provide the necessary anti-aliasing filtering. This book introduces the concepts of analog multirate signal processing for the efficient implementation of 2-D filtering in integrated circuit form, particularly from the viewpoints of silicon area and power dissipation.

After reviewing the basic concepts of 2-D signals and filter systems, efficient 2-D SC networks and design techniques are described to implement 2-D decimating filters, both with finite impulse response (FIR) and infinite impulse response (IIR) with separable denominator polynomial. Such techniques offer simple, systematic synthesis procedures which significantly improve the more traditional available design techniques for 2-D multirate filters.

A 2-D SC image processor that realizes both (2 x 2)nd-order Butterworth lowpass and highpass filtering functions for video image signals is demonstrated as a prototype integrated circuit implemented in 1.0-μm CMOS technology. It employs a 2-fold decimating horizontal filter to reduce the size of the DL memory blocks and increase their access time. Careful analysis of speed limitation effects associated with the write operation mode of the storage cells allowed to design only one simple type of operational transconductance amplifier for both the vertical filter and associated DL memory blocks and the horizontal filter. Because of the 2-fold horizontal decimation the total number of storage capacitors is only 2,280 compared to 4,560 that would be needed in a non-multirate system. The experimental characterization of this prototype chip demonstrated the feasibility of real-time multirate 2-D image processing with equivalent 8-bits accuracy, using only 2.5 x 3.0 mm^2 of silicon area and dissipating as little as 85 mW at 5 V supply and 18 MHz sampling rate.

1

INTRODUCTION

1.1 INTRODUCTORY REMARKS

A Two-Dimensional (2-D) signal can be represented by a frequency spectrum that can be modified, reshaped, or manipulated through 2-D signal processing techniques. Image processing is a particular case of 2-D signal processing whereby a 2-D image is transformed in order to extract useful information to improve its visual effect or to change it into some more suitable form for further analysis. In robotics, for example, an edge detection vision system is often required because the determination of the geometric shape of the object to be handled by the robot is critical to successful robot motion. In this case, a 2-D highpass filter can be used to strengthen the contour of the object. Besides detection vision systems, 2-D signal processing techniques have many other important applications, such as remote sensing via satellite, radar, sonar and acoustic image processing, medical and seismic data processing, video signal compression and communication, video special effects and even the planning of power systems [1.1 – 1.9].

Similarly to their one-dimensional (1-D) counterparts, 2-D discrete-time filters can have either finite impulse response (FIR) or infinite impulse response (IIR) transfer functions. 2-D FIR filters have three important advantages over their IIR counterparts. First, they are always stable and hence the application of complex stability tests is unnecessary [1.1], [1.3]. Second, a linear phase response with respect to both the horizontal and vertical frequency variables can easily be achieved. Third, FIR filters can be efficiently implemented using fast processing algorithms such as the Fast Fourier Transform (FFT) [1.1], [1.3]. Although a linear phase response is particularly important in the processing of images, the selectivity that can be achieved with these filters is limited. Consequently, in applications where rapid transitions are required between passbands and stopbands, large filter orders would be required to meet the specifications. As a result, the amount of computation involved in the application of these filters can be prohibitively high. This problem can be overcome by using instead 2-D IIR filters. In applications where linear phase responses are required, 2-D IIR filters can also be designed with quasi-linear phase response [1.2 – 1.3].

2-D filters can be realized in either digital or analog form. Digital 2-D filters are most commonly used in modern image processing applications [1.10 – 1.12] because of the well-known advantages regarding flexible VLSI design and implementation, functional programmability, high accuracy, low noise and distortion, and high processing speed. In addition, digitized image signals can be further processed and transmitted with high quality. Despite their advantages, there are many applications of digital 2-D filters when analog-to-digital (A/D) and digital-to-analog (D/A) converters are required to convert the signals between an analog image source and the digital processing core. Usually, the addition of on-chip A/D and D/A converters leads to increased power dissipation, silicon area and may even introduce extra noise and distortion that might decrease the overall quality of image processing.

By contrast with their digital counterparts, analog implementations of 2-D filters are particularly relevant for applications where only moderate processing accuracy is needed but small chip area and low power dissipation are the primary factors of concern [1.9], [1.13 – 1.24]. Examples of these can be found in low-cost applications of mobile robotics vision systems where a 2-D edge enhancement processing function is usually needed [1.3], [1.18]. Analog 2-D filters can be realized using either continuous-time or discrete-time circuits. In the continuous-time domain there have been circuit realizations of 2-D filters based on traditional analog signal processing techniques [1.13], [1.25] as well as circuit realizations based on neural network techniques [1.9], [1.17 – 1.22]. The realization of analog 2-D discrete-time filters has also been investigated using switched-capacitor (SC) circuits [1.14 – 1.16], [1.24], [1.26].

Regardless of which form of realization, however, considerable differences exist between 2-D filtering networks and the more common 1-D filtering networks. One such major difference is the amount of data involved in typical applications. In speech processing, for instance, speech is typically sampled at a 10 kHz rate and thus yielding 10 thousand data points to process in a second. In video processing, by contrast, we may have 25 frames/s with one frame consisting of 625 x 1140 pixels. Thus, in this case, there will be about 18 million data points to be processed per second, which is many orders of magnitude greater than in the case of speech processing. Due to this difference in data rate requirements, the computational and implementation efficiency of the processing algorithms plays a rather important role in 2-D signal processing, either analog or digital.

The second major difference between the VLSI implementation of 2-D and 1-D filtering networks is the required silicon area. Image filtering requires the use of delay-line (DL) memory blocks each of which may contain about 1,000 unit-delays or storage cells. A storage cell is generally built using one capacitor and several transistors. Typically, the size of the DL memory block is related to the 2-D signal sampling frequencies. For example, a video signal where the horizontal and vertical sampling frequencies, respectively, are $F_{s2} = 17.718$ MHz and $F_{s1} = 15.625$ kHz, leads to a total of 1,140 storage cells needed in one DL memory block to memorize a full line-segment. To realize a high order 2-D filter several of such large DL memory blocks have to be employed, which not only occupies silicon area but also increases the overall power dissipation in both digital and analog circuits. In the analog implementation, in particular, this may become even more important because building large analog memory blocks will add noise and distortion [1.16], [1.20], [1.22 – 1.24].

One particular problem related to the design of 2-D analog filters concerns the lack of direct design methodologies employing mixed-signal (s, z)-variables. As a result, the filter design has to be converted first to the 2-D discrete-time domain by using the well-known s-to-z transformations and then, after designing the model digital filter, the inverse z-to-s transformation is needed to convert back to the s domain and thus obtaining the desired analog filter response [1.13 – 1.15], [1.25 – 1.26].

The use of 2-D multirate digital processing techniques can significantly reduce the size of the required DL memory blocks [1.27] and have therefore been used for such applications as image subband-filtering for coding/decoding, sampling format conversion and de-interlacing between various video standards and HDTV systems [1.27 – 1.31]. However, the current existing 2-D multirate techniques cannot be efficiently applied to the 2-D analog filtering domain because there are no reported efficient design techniques for 2-D IIR multirate digital filters. This book describes new design techniques for efficient implementations of 2-D analog multirate filters both with FIR and IIR filtering functions. They are based on extensions of conventional 1-D multirate SC polyphase networks and offer simple, systematic synthesis procedures which significantly extend the current available design techniques for 2-D multirate filters.

1.2 BOOK OUTLINE

Besides this introductory chapter, the book is divided into five additional chapters. Chapter 2 reviews the basic concepts of 2-D signals and filter systems, both analog and discrete-time. Given their relevance in the context of this work 2-D discrete-time filtering systems are examined in more detail, both with FIR and IIR transfer functions. Then, after discussing the major comparative advantages of analog and digital filter VLSI implementations, the chapter goes on to discuss the key aspects of implementing very large DL memory blocks. At the end of the chapter, various 2-D signal filtering applications are described for illustration purposes, indicating noise removal, edge enhancement, seismic signal processing, HDTV subband filtering and video image special effects.

Chapter 3 describes the fundamental aspects of 2-D signal multirate filtering. First, it reviews the basic concepts of 1-D signal decimation filtering and efficient polyphase implementations, and then considers also the case of 2-D decimation filtering. In particular, it examines the requirements of the critical DL memory blocks and shows how multirate techniques can be effectively employed to reduce their size and thus lead to much more efficient 2-D filter implementations then their non-multirate counterparts.

Chapter 4 introduces a new type of 2-D decimation filter structures that are designated as polyphase-coefficient structures. To this end, it first presents the modified 1-D decimation polyphase filter structures [1.32 – 1.38] and shows how alternative expressions of the decimation filters can be derived leading to the proposed polyphase-coefficient structures, both for FIR and IIR filters [1.39]. Then, these results are readily extended to obtain also polyphase-coefficient structures for 2-D decimation FIR and IIR filters.

Chapter 5 presents SC architectures for implementing the 2-D decimation filters in polyphase-coefficient form [1.40 – 1.42]. First, the most commonly used SC building blocks for multirate signal processing are reviewed. Then, four different cases are considered to describe the 2-D SC polyphase-coefficient implementation as a function of the decimating factors and filtering order in both horizontal and vertical dimensions. Design examples for both FIR and IIR 2-D SC decimation filter are given to illustrate the derivation of the proposed SC circuits.

Finally, Chapter 6 demonstrates the design and implementation of a real-time 2-D analog multirate image processor prototype chip realized in 1.0-μm CMOS technology [1.43 – 1.44]. The most relevant design aspects are discussed for both the SC horizontal decimation filter and SC vertical filter with associated DL memory blocks. The evaluation of major parasitic effects in the analog DL memory block is carefully considered to allow designing the operational transconductance amplifiers for minimum power dissipation of the overall system. A detailed chip experimental characterization is presented, including the demonstration of its operation for real-time video highpass filtering.

REFERENCES

[1.1] J. S. Lim and A. V. Oppenheim, *"Advanced Topics in Signal Processing"*, Prentice Hall, 1988.

[1.2] P.P. Vaidyanathan, *"Multirate Systems and Filter Banks"*, Prentice-Hall, Englewood Cliffs, NJ, 1993.

[1.3] Wu-Sheng Lu and Andreas Antoniou, *"Two-Dimensional Digital Filters"*, Marcel Dekker, Inc., 1992.

[1.4] M.Soumekh, *"Fourier Array Imaging"*, Prentice Hall, 1994.

[1.5] R.C. Gonzalez and Paul Wintz, *"Digital Image Processing"*, Addison-Wesley Publishing Company, 1977.

[1.6] B. Ackland, "VLSI Architectures for Multimedia and Video Conferencing", *Proc. IEEE International Symposium on Circuits and Systems*, London, pp. 147, Chapter 3.21, Tutorial, 1994.

[1.7] T. Fujio, "HDTV Systems", *Proceedings. of the IEEE*, Vol. 73, No. 4, pp. 646-655, April, 1985.

[1.8] C. Ngo, "Image Resizing and Enhanced Digital Video Compression", *EDN*, pp. 145-155, January 1996.

[1.9] C. Mead, *"Analog VLSI and Neural Systems"*, Addison-Wesley Publishing Inc., 1989.

[1.10] R.B. Yates, N.A. Thacker, S.J. Evans, S.N. Walker and P.A. Ivey, "An Array Processor for General Purpose Digital Image Compression", *IEEE Journal of Solid-State Circuits*, Vol. 30, No. 3, pp. 1096-1101, March 1995.

[1.11] C. Joanblanq, P. Senn and M.J. Colaitis, "A 54 MHz CMOS Programmable Video Signal Processor for HDTV Applications", *IEEE Journal of Solid-State Circuits,* Vol. 25, No. 3, pp. 730-734, June 1990.

[1.12] C.J. Kulach, L.T. Bruton and N.R. Bartley, "A Real-Time Video Implementation of a 3-D 1st-Order Recursive Discrete-Time Filter", *IEEE International Symposium on Circuits and Systems*, pp. 609-612, Atlanta, GA., May 1996.

[1.13] M. A. Sid-Ahmed, "Two-Dimensional Analog Filters: A New Form of Realization", *IEEE Transactions on Circuits and Systems*, vol. 36, no. 1, pp. 153-154, January 1989.

[1.14] K. Nishikawa, T. Takebe and M. Hayashihara, "Two-Dimensional Switched-Capacitor Filter", *IEEE International Symposium on Circuits and Systems*, pp. 73-76, Montreal, Canada, June 1984.

[1.15] A. Handkiewicz, "Two-Dimensional Switched-Capacitor Filter Design System for Real-Time Image Processing", *IEEE Transactions on Circuits and Systems for Video Technology*, vol. 1, no.3, pp. 241-246, September 1991.

[1.16] K. A. Nishimura and P. R. Gray, "A Monolithic Analog Video Comb Filter in 1.2-µm CMOS", *IEEE Journal of Solid-State Circuits*, Vol. 28, No. 12, pp. 1331-1339, December 1993.

[1.17] V. Ward and M. Syrzycki, "VLSI Implementation of Receptive Fields with Current-Mode Signal Processing for Smart Vision Sensors", *Analog Integrated Circuits and Signal Processing*, vol. 26, no.6, pp. 167-179, 1995.

[1.18] P.Kinget and M.S.J.Steyaert, "A Programmable Analog Cellular Neural Network CMOS Chip for High Speed Image Processing", *IEEE Journal of Solid-State Circuits*, vol. 30, no.3, pp. 235-243, March 1995.

[1.19] A.P. Almeida, "A Low-Power Implementation of an Analog Retina Model", *GCSI Group Internal Report*, Instituto Superior Técnico, March 1995.

[1.20] E.A. Vittoz, "Analog VLSI Signal Processing: Why, Where, and How?", *Analog Integrated Circuits and Signal Processing*, No. 6, pp. 27-44, May 1994.

[1.21] D. Gerna, M. Brattoli, E. Chioffi, G. Colli, M. Pasotti and A. Tomasini, "An Analog Memory for a QCIF Format Image Frame Storage", *IEEE International Symposium on Circuits and Systems*, pp. 289-292, Atlanta, May 1996.

[1.22] C.A. Mead and T. Delbruck, "Scanners for Visualizing Activity of Analog VLSI Circuit", *Analog Integrated Circuits and Signal Processing*, No. 1, pp. 93-106, June 1991.

[1.23] E. Franchi, M. Tartagni, R. Guerrieri and G. Baccarani, "Random Access Analog Memory for Early Vision", *IEEE Journal of Solid-State Circuits*, Vol. 27, No. 7, pp. 1105-1109, July 1992.

[1.24] K.Matsui, T.Matsuura, S.Fukasawa, Y.Izawa, Y.Toba, N.Miyake and K.Nagasawa, "CMOS Video Filters Using SC 14-MHz Circuits", *IEEE Journal of Solid-State Circuits*, Vol. 20, No. 6, pp. 1096-1101, December 1985.

[1.25] H. J. Kaufman and M. A. Sid-Ahmed, "2-D Analog Filters for Real -Time Video Signal Processing", *IEEE Transactions on Consumer Electronics*, vol. 36, no.2, pp. 137-140, May 1990.

[1.26] A. Handkiewicz, "Two-Dimensional SC Filter Design Using a Gyrator-Capacitor Prototype", *International Journal of Circuit Theory and Applications*, Vol. 16, pp. 101-105, 1988.

[1.27] U. Pestel and K. Gruger, "Design of HDTV Subband Filterbanks Considering VLSI Implementation Constraints", *IEEE Transactions on Circuits and Systems for Video Technology*, vol. 1, no.1, pp. 14-21, March 1991.

[1.28] R. Ansari and C.L. Lau, "2-D IIR Filters for Exact Reconstruction in Tree-Structured Sub-Band Decomposition", *Electronics Letters*, vol. 23, No.12, pp. 633-634, 4th June 1987.

[1.29] Q.S. Gu, M.N.S. Swamy, L.C.K. Lee and M.O. Ahmad, "IIR Digital Filters for Sampling Structure Conversion and Deinterlacing of Video Signals", *IEEE International Symposium on Circuits and Systems*, pp. 973-976, San Diego, May 1995.

[1.30] G. Schamel, "Pre- and Post-filtering of HDTV Signals for Sampling Rate Reduction and Display Up-conversion", *IEEE Transactions on Circuits and Systems*, vol. 34, no.11, pp. 1432-1439, November 1987.

[1.31] P. Siohan, "2-D FIR Filter Design for Sampling Structure Conversion", *IEEE Transactions on Circuits and Systems for Video Technology*, vol. 1, no.4, pp. 337-350, December 1991.

[1.32] R. Gregorian and W.E. Nicholson, "SC Decimation and Interpolation Circuits", *IEEE Transactions on Circuits Systems*, vol. 27, pp. June 1980.

[1.33] D.C. Grunigen, U.W. Brugger and G.S. Moschytz, "A Simple SC Decimation Circuit", *Electronics Letters*, vol. 17, pp. January 1981.

[1.34] D.C. Grunigen, R. Sigg, M. Ludwig, U.W. Brugger, G.S. Moschytz and H. Melchior, "Integrated SC Low-Pass Filter with Combined Anti-Aliasing Decimation Filter for Low Frequencies", *IEEE Journal of Solid-State Circuits*, Vol. 17, pp. 1024-1028, December 1982.

[1.35] J. E. Franca, "Non Recursive Polyphase Switched-Capacitor Decimators and Interpolators", *IEEE Transactions on Circuits and Systems*, vol. CAS-32, no. 9, pp. 877-887, September 1985.

[1.36] J. E. Franca, "SC Systems for Narrow Bandpass Filtering", *PhD Thesis,* Imperial College, University of London, 1985.

[1.37] J. E. Franca and S. Santos, "FIR Switched-Capacitor Decimators with Active-Delayed Block Polyphase Structures", *IEEE Transactions on Circuits Systems*, vol. 35, no. 8, pp. 1033-1037, August 1988.

[1.38] R.P. Martins and J.E. Franca, "A 2.4-µm CMOS SC Video Decimator with Sampling Rate Reduction from 40.5 MHz to 13.5 MHz", *IEEE Custom Integrated Circuits Conference,* pp. 1-4, San Diego, May 1989.

[1.39] W. Ping and J. E. Franca, "New Form of realization of IIR SC decimators", *Electronics Letters*, vol. 29, no. 11, pp. 953-954, May 27, 1993.

[1.40] W. Ping and J. E. Franca, "SC Polyphase Structures for 2-D Analog FIR Filtering", *IEEE International Symposium on Circuits and Systems*, pp. 1038-1041, Chicago, May 1993.

[1.41] W. Ping and J. E. Franca, "SC Decimation Techniques for 2-D IIR Filtering", *IEEE Visual Signal Processing and Communication*, pp. 223-226, Melbourne, 1993.

[1.42] W.Ping and J.E.Franca, "2-D SC Decimating Filters", *IEEE Transactions on Circuits and Systems - Part I,* Vol. 43, No. 4, pp. 257-270, April 1996.

[1.43] W. Ping and J. E. Franca, "A Very Compact 1.0-µm CMOS SC Multirate 2-D Image Filter", *Proc. of European Solid-State Circuits Conference,* Switzerland, September 1996.

[1.44] W. Ping and J. E. Franca, "A CMOS 1.0-µm 2-D Analog Multirate System for Real-Time Image Processing", *IEEE Journal of Solid-State Circuits,* Vol. 32, No. 7, pp. 1037-1048, July, 1997.

2

2-D SIGNALS AND FILTERING SYSTEMS

2.1 INTRODUCTION

A two-dimensional (2-D) analog signal can be represented as a function $x(t_1, t_2)$ of two independent continuous-time variables, t_1 and t_2, whereas a 2-D analog image signal is represented as a function $x(t_2, n_1 T_1)$ of one continuous-time variable, t_2, and one discrete-time variable, n_1. Sampling of a 2-D analog image signal produces a 2-D discrete-time image signal represented as a function $x(n_2, n_1)$ of two discrete-time variables, n_2 and n_1.

One major difference between 2-D signals and the more familiar 1-D signals reside in the much larger amount of data involved for the processing of the former in comparison to the latter. For example, in speech processing, which is an important application area of 1-D signal processing, speech is typically sampled at 10 kHz and hence there are 10,000 data points to process in a second. In video processing, by contrast, which is an important application area for 2-D signal processing, we may typically have a frame processing speed of 25 frames per second where each frame may typically contain 625 lines and each line 1140 pixels. This gives a total of about 18,000,000 data points to process per second, which is many orders of magnitude greater than in the case of speech processing. This is a fundamental reason why implementation efficiency is paramount for the successful realization of 2-D filtering systems, both analog and digital, in integrated circuit form [2.1 – 2.11].

2.2 2-D SIGNALS

2.2.1 Continuous-Time Line Signals

A 2-D analog signal is a physical quantity that depends on two independent variables, t_1 and t_2. It can be represented by a function $x(t_1, t_2)$ where each of the two variables t_1 and t_2 may represent time, distance, or any other physical variable. For example, the light intensity of an image is a 2-D continuous-time signal represented as a function of distance in the x and y directions.

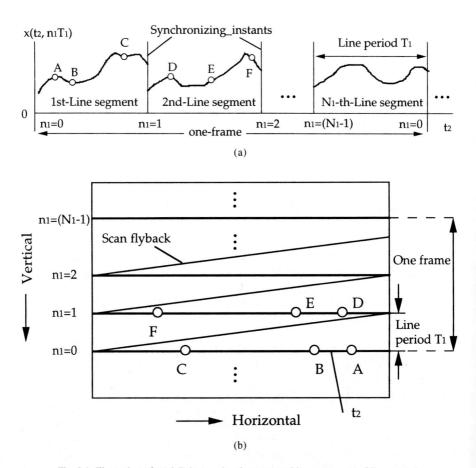

Fig. 2.1. Illustration of (a) 2-D image signal represented by a sequence of lines and (b) a frame representation of the image in the t, nT space.

A general 2-D analog *image* signal can be represented by a sequence of line segments and image frames, as illustrated in Fig. 2.1. Each line segment, shown in

Fig. 2.1(a), contains a series of continuous-time pixels and each image frame, shown in Fig. 2.1(b), contains a number of lines. Such signal can be generated, for example, by a television camera focused on a scene. An image of the scene is formed on a light sensitive surface in the camera so as to produce an electrical charge pattern. The amount of charge at each point depends on the luminance of the corresponding element of the scene. The charge patterns are converted into currents which, in turn, are read out by means of the raster-type electron beam scanning scheme depicted in Fig. 2.1(b). The scanning of the electron beam is achieved by simultaneous horizontal and vertical deflections of the beam. The horizontal motion is called the *line scan*, and each segment of the analog signal represents a *line signal*.

Mathematically, the above 2-D signal can be expressed as a function $x(t_2, n_1 T_1)$ of a continuous variable t_2 and a discrete variable $n_1 T_1$. An abbreviated representation is written as

(2.1) ... $\qquad x(t_2, n_1) = x(t_2, n_1 T_1)$

where t_2 represents the time in the horizontal direction in each line scan, n_1 represents the vertical discrete-time variable and T_1 can be considered the sampling time period along the vertical direction.

Filtering of a 2-D image signal can be viewed as the process whereby different points of the same line segment and of different lines are correlated, as illustrated in Fig. 2.1. Horizontal filtering, on the one hand, corresponds to evaluating only the horizontal signal on each line scan correlation, for example C with B, A and then F with E, D. Vertical filtering, on the other hand, corresponds to the evaluation of the points in the vertical direction, for example D with A, E with B and F with C. Hence, the overall filtering effect in both horizontal and vertical dimensions is obtained by the combined evaluation of all signal points A, B, C, D, E and F.

The most familiar 2-D image signal is the broadcast television signal. There are several television standards around the world, the well known of which are the NTSC system, SECAM system and PAL system [2.7], [2.12 – 2.13]. In the PAL system, in particular, the image signal contains 25 frames per second, 625 lines per frame interlaced by 2 fields, and typically 1140 pixels per line. In general, the signal can be represented by the luminance signal (Y) and the chrominance signals (I, Q). The Y signal contains most of the video information and it requires a bandwidth of about 5 MHz. The I and Q signals contain color information and they require a much smaller bandwidth of about 1 MHz. In general, image filtering corresponds to the processing of the luminance signal. Thus, throughout this dissertation, the function $x(t_2, n_1)$ represents the luminance signal.

The Fourier transform of a 2-D signal gives the frequency spectrum

(2.2) ... $\qquad X(\Omega_2, \omega_1)$,

where Ω_2 is the horizontal frequency variable corresponding to the continuous-time variable t_2 and ω_1 is the vertical frequency variable corresponding to the discrete-time variable n_1. For example, a 2-D image signal spectrum can be illustrated by the three-dimensional (3-D) surface shown in Fig. 2.2(a), and whose topview 2-D form shown in Fig. 2.2(b) is also used to represent the frequency spectrum. It must be noted that, for convenience of representation, such 3-D surfaces are often drawn simply in topview 2-D form.

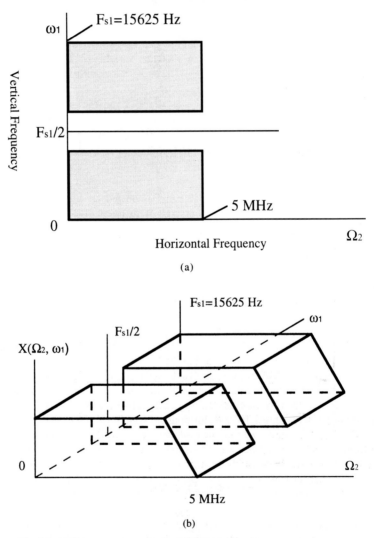

Fig. 2.2. (a) Frequency spectrum of a 2-D signal in 3-D form and (b) its topview representation.

2.2.2 Discrete-Time Line Signals

Similarly to the one-dimensional (1-D) sampling theorem, the 2-D signal sampling theorem states that a band-limited 2-D continuous-time signal $x(t_2, n_1T_1)$, i.e. a 2-D continuous-time signal whose frequency spectrum is zero for all frequencies (Ω_2, ω_1) such that $|\Omega_2| \geq \omega_{s2}/2$ or $|\omega_1| \geq \omega_{s1}/2$, sampled at the sampling frequencies $\omega_{s2} = 2\pi/T_2$ and $\omega_{s1} = 2\pi/T_1$ can be uniquely determined from the resulting sampled values $x(n_2T_2, n_1T_1)$.

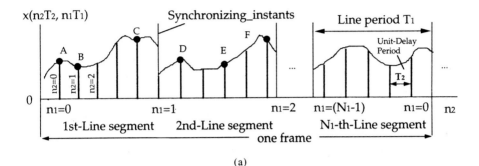

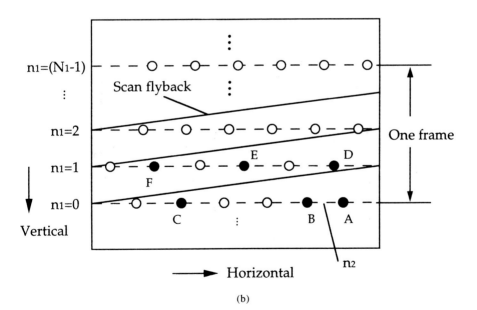

Fig. 2.3. (a) 2-D sampled image signal for each line scan and (b) a frame representation of the image.

Here, we should recall from the previous discussion that the 2-D image signal $x(t_2, n_1T_1)$ is inherently sampled in vertical dimension and therefore the 2-D sampling process occurs only in horizontal dimension. Thus, a sampled 2-D signal can be represented in abbreviated form by

(2.3) ... $\quad x(n_2, n_1) = x(n_2T_2, n_1T_1)$

An illustrative representation of a sampled 2-D signal can be viewed in Fig. 2.3.

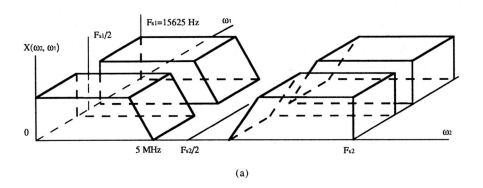

(a)

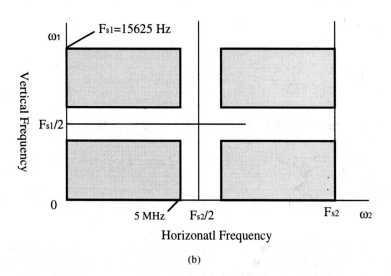

(b)

Fig. 2.4. (a) Illustration of the frequency spectrum of the 2-D signal in Fig. 2.3. (b) Topview version of (a).

Similarly to the 2-D analog case discussed before, 2-D discrete-time filtering corresponds to evaluating together the relationships among discrete-time points of a sampled 2-D signal, e.g. A, B, C, D, E and F in Fig. 2.3. The 2-D frequency spectrum of such signal can be obtained by applying a 2-D discrete-time Fourier transform yielding

$$(2.4) \quad X(\omega_1, \omega_2) = \sum_{n_1=-\infty}^{\infty} \sum_{n_2=-\infty}^{\infty} x(n_1, n_2) e^{-j\omega_2 n_2} e^{-j\omega_1 n_1}$$

where ω_2 is the horizontal frequency variable corresponding to n_2, and ω_1 is the vertical radial frequency variable corresponding to n_2. Fig. 2.4 illustrates a 2-D discrete-time frequency spectrum where F_{s2} and F_{s1}, respectively, are the sampling frequencies of the horizontal and vertical dimensions.

In addition, we can also define the 2-D z-transform $X(z_2, z_1)$ of a signal sequence $x(n_2, n_1)$ such that

$$(2.5) \quad X(z_1, z_2) = \sum_{n_1=-\infty}^{\infty} \sum_{n_2=-\infty}^{\infty} x(n_1, n_2) z_2^{-n_2} z_1^{-n_1}$$

The 2-D discrete-time Fourier transform given in (2.4) can also be derived from the above expression by using the complex representations of $z_1 = r_1 e^{j\omega_{s1}}$ and $z_2 = r_2 e^{j\omega_{s2}}$.

2.3 2-D IMAGE FILTER SYSTEMS

2.3.1 Processing of Continuous-Time Line Signals

Similarly to 1-D filtering, a 2-D signal can be represented by a frequency spectrum that can be modified, reshaped, or manipulated through 2-D filtering. This type of processing can be carried-out by using 2-D filters which allow filtering the image signal not only along each horizontal line scan but also in the vertical direction.

A linear, shift-invariant and causal 2-D filter can be characterized either in terms of difference equations or state-space equations *in two independent variables*, or in terms of transfer functions or matrices of transfer functions, and which are rational functions of polynomials in two variables. The time-domain analysis in 1-D filters is replaced by *space-domain* analysis in 2-D filters and the frequency-domain analysis is referred to as *spatial-frequency* to emphasize that frequency may not bear an inverse relation with time. The transfer function of a 2-D filter yields amplitude and phase responses represented by 3-D surfaces over a 2-D frequency plane, rather than curves plotted over a frequency axis as is the case in 1-D filters.

As previously mentioned, when the line signals are continuous-time an analog image signal is represented by a function $x(t_2, n_1)$ where t_2 is a continuous-time variable and n_1 represents a discrete-time variable associated with the vertical sampling period T_1. Let a 2-D image signal with continuous-time lines be processed by an analog filter with impulse response $h(t_2, n_1)$, as represented in Fig. 2.5.

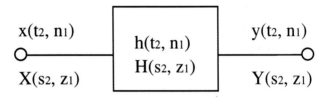

Fig. 2.5. 2-D analog filter with continuous-time line processing.

Such a linear, time-invariant system can be expressed in terms of the difference equation [2.3], [2.11]

$$(2.6) \quad y(t_2, n_1) = \sum_{k_1=0}^{\infty} \int_{\alpha_2=0}^{\infty} h(\alpha_2, k_1)\, x(t_2 - \alpha_2, n_1 - k_1)\, d\alpha_2,$$

where a_2 and k_1 are, respectively, the convolution intermediate variables in both dimensions. The filter transfer function $H(s_2, z_1)$ can be expressed as

$$(2.7) \quad H(s_2, z_1) = \frac{\displaystyle\sum_{i_1=0}^{N_1-1}\sum_{i_2=0}^{N_2-1} a_{i_1,i_2}\, s_2^{-i_2} z_1^{-i_1}}{1 + \displaystyle\sum_{\substack{i_1=0 \\ i_1+i_2 \neq 0}}^{N_1-1}\sum_{i_2=0}^{N_2-1} b_{i_1,i_2}\, s_2^{-i_2} z_1^{-i_1}} = \frac{Y(s_2, z_1)}{X(s_2, z_1)},$$

where $(N_2 \times N_1)$ gives the order of the 2-D filter, $a_{i1,i2}$ and $b_{i1,i2}$ are the 2-D filter coefficients, s_2 is the Laplace-variable representing the horizontal frequency and z_1^{-1}

is the unit-delay term that is referred to as the delay-line (DL) memory in the vertical dimension.

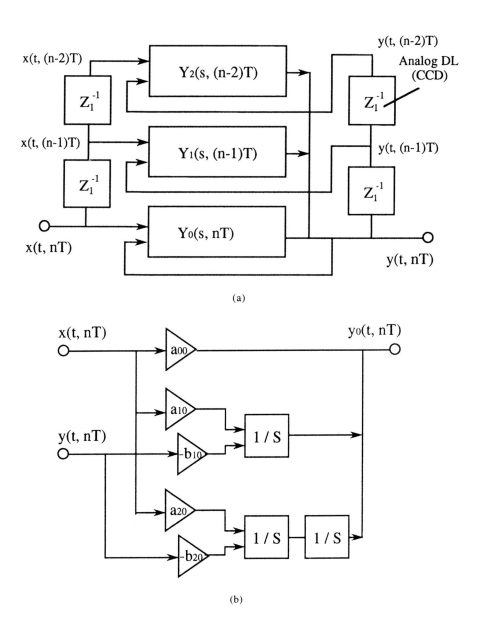

Fig. 2.6. Direct-form realization of a (2 x 2)nd-order analog filter. (a) Block-diagram of the overall 2-D analog filter and (b) possible realization of $Y_0(s, nT)$.

In order to design a 2-D filter represented by the transfer function above, where both continuous-time (s) and discrete-time (z) variables co-exist, it is first necessary to transform the s variable of the above expression into the corresponding z variable using appropriate s-to-z transformations. This can thus be fully designed in the discrete-time domain. Then, by applying the corresponding z-to-s transformation to convert z back to the s domain we finally achieve the desired filter $H(s_2, z_1)$. A detailed discussion of such design issues, which are outside the scope of this book, can be found in a large body of literature [2.9], [2.11], [2.14 – 2.17].

The implementation structures for the above types of analog filters are closely related to their 2-D digital counterparts since $H(s_2, z_1)$ contains a discrete-time variable n_1. For example, the block diagram of Fig. 2.6(a) illustrates the direct-form implementation of a (2 × 2)nd-order filter. It uses a simplified representation such as the variable t_2 is replaced by t and n_2 by n. $Y_0(s, nT)$ is a function of only the nth scanning line signal whereas $Y_1(s, (n-1)T)$ is a function of only the $(n-1)$ scanning line signal, and so forth. A possible realization of $Y_0(s, nT)$ is schematically illustrated in Fig. 2.6(b).

2.3.2 Processing Discrete-Time Line Signals

We have seen before that a 2-D image signal with time-discrete line segments can be represented by discrete-time functions $x(n_2, n_1)$. To process such signals 2-D discrete-time filters are required and which, similarly to their 1-D counterparts [2.18 – 2.20], can be realized with either finite impulse response (FIR) or infinite impulse response (IIR). FIR filters possess three important advantages over IIR filters. First, they are always stable and hence the application of complicated stability tests needed in IIR filters is unnecessary. Second, a linear phase response with respect to the frequency variables ω_1 and ω_2 can be easily obtained, and which is particularly important in image processing applications [2.1 – 2.5]. Third, owing to their finite impulse response, these filters can be implemented in terms of FFT.

Consider a linear, time-invariant 2-D FIR filter system characterized by the transfer function

$$(2.8) \quad H(z_1, z_2) = \sum_{n_1=0}^{N_1-1} \sum_{n_2=0}^{N_2-1} h(n_1, n_2) z_2^{-n_2} z_1^{-n_1},$$

where N_1 and N_2, respectively, represent the filter length in the z_1 and the z_2 dimensions. The z_1^{-1} delay term, representing a DL memory block, refers to the sampling frequency $M_1 F_{s1}$ along the vertical axis, whereas the z_2^{-1} delay term refers to the sampling frequency $M_2 F_{s2}$ along the horizontal axis. Such 2-D discrete-time filter with impulse response $h(n_2, n_1)$ is represented in Fig. 2.7.

2.3.2.1 *2-D FIR Filtering*

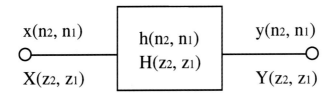

Fig. 2.7. 2-D discrete-time filter with discrete-time line processing.

When the transfer function $H(z_2, z_1)$ is separable, and hence can be expressed as

$$(2.9) \ldots H(z_1, z_2) = \left(\sum_{n_1=0}^{N_1-1} h_1(n_1) z_1^{-n_1} \right) \left(\sum_{n_2=0}^{N_2-1} h_2(n_2) z_2^{-n_2} \right) = H_1(z_1) H_2(z_2)$$

the system function can be regarded as the multiplication of the transfer functions of two simpler 1-D digital filters. Such separability is of considerable practical importance since it allows the design and realization of a 2-D digital filter to be carried out by two cascaded 1-D filters, as illustrated in the block diagram of Fig. 2.8, and which reduces the complexity of design and hardware implementation [2.2], [2.21 – 2.24]. The frequency response of the filter can be readily obtained by appropriately combining the frequency responses of each one of the 1-D filters.

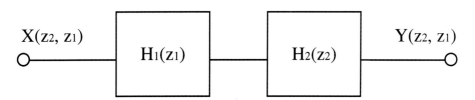

Fig. 2.8. Block diagram of a separable 2-D digital filter.

The most relevant design techniques of 2-D FIR filters, e.g. windowing, frequency sampling, optimal filter design and transformation, are described in detail in references [2.1 – 2.2], [2.4 –2.5], [2.23 – 2.25]. The various structures available for implementing such filters, e.g. direct, parallel, cascade and separable structures, can vary significantly in their computational complexity, sensitivity to coefficient quantization, level of output noise produced by the quantization of products, and processing speed. For the general transfer function (2.8), Fig. 2.9 illustrates a direct-form architecture where z_2^{-1} represents a unit-delay in the horizontal processing dimension, z_1^{-1} represents a line-delay in the vertical processing dimension and the arrows represent the coefficient-multipliers weighted with the corresponding impulse response coefficients of the filter.

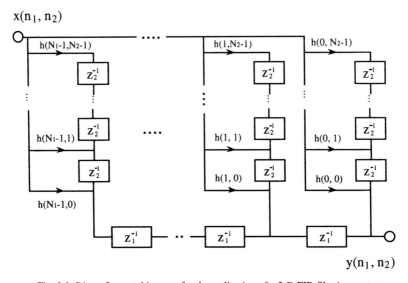

Fig. 2.9. Direct-form architecture for the realization of a 2-D FIR filtering system.

From the figure above, we can further notice that the 2-D FIR transfer function can be decomposed as a 1-D non-recursive filter in z_1 and whose coefficients are, in turn, a function of z_2. Therefore, for its implementation, classical 1-D direct-form structures can be employed in both dimensions.

2.3.2.2 *2-D IIR Filtering*

Although 2-D FIR filters are always stable and can be readily designed to produce a linear phase response, the level of selectivity that can be achieved is rather limited. Hence, in applications where high selectivity is needed high filter orders would be required and which, in turn, could render the amount of computation involved prohibitively high. This problem can be overcome by using instead 2-D IIR filters [2.2], [2.12], [2.21], [2.22], [2.26 – 2.27]. In applications where linear phase responses are required, such 2-D IIR filters can be designed with a quasi-linear phase response to approximate the linear phase response of FIR filters [2.26].

In general, a linear, shift-invariant IIR filter can be described by

$$(2.10) \ldots H(z_1, z_2) = \frac{N(z_1, z_2)}{D(z_1, z_2)} = \frac{\sum_{n_1=0}^{N_1-1} \sum_{n_2=0}^{N_2-1} a_{n_1, n_2} z_2^{-n_2} z_1^{-n_1}}{1 + \sum_{\substack{n_1=0 \\ n_1+n_2 \neq 0}}^{N_1-1} \sum_{n_2=0}^{N_2-1} b_{n_1, n_2} z_2^{-n_2} z_1^{-n_1}}$$

where, as before, z_2^{-1} and z_1^{-1} are, respectively, the unit-delay terms referring to the sampling frequencies $M_2 F_{s2}$ and $M_1 F_{s1}$.

Unlike in the case of 1-D filters considered before, the implementation of the 2-D transfer functions are not generally applicable to all types of architectures. First, since a 2-D polynomial can not, in general, be factored as a product of lower-order polynomials, $H(z_2, z_1)$. can not be factored either as a product of lower-order polynomials. Therefore, cascade architectures are not generally adequate for the implementation of 2-D filters. Because of the non-factorability of a 2-D polynomial, ensuring the stability of a 2-D IIR filter is usually a task which may impose considerable restrictions for designing and implementation [2.1 – 2.2].

Due to the above constraints for implementation, there are usually two cases considered for the implementation of 2-D IIR filtering functions. In one case only the denominator polynomial is separable, such that the transfer function can be expressed by

$$(2.11) \ldots H(z_1, z_2) = \frac{\sum_{n_1=0}^{N_1-1} \sum_{n_2=0}^{N_2-1} a_{n_1, n_2} z_2^{-n_2} z_1^{-n_1}}{(1 + \sum_{n_1=1}^{N_1-1} b_{n_1} z_1^{-n_1})(1 + \sum_{n_2=1}^{N_2-1} b_{n_2} z_2^{-n_2})} = \frac{N(z_1, z_2)}{D_1(z_1) D_2(z_2)}.$$

The other case is when both the denominator and nominator polynomials are separable, and hence the transfer function can be given by

$$H(z_1,z_2) = \frac{(\sum_{n_1=0}^{N_1-1} a_{n_1} z_1^{-n_1})(\sum_{n_2=0}^{N_2-1} a_{n_2} z_2^{-n_2})}{(1+\sum_{n_1=1}^{N_1-1} b_{n_1} z_1^{-n_1})(1+\sum_{n_2=1}^{N_2-1} b_{n_2} z_2^{-n_2})}$$

(2.12)...

$$= \frac{N_1(z_1)N_2(z_2)}{D_1(z_1)D_2(z_2)} = H_1(z_1)H_2(z_2)$$

In this case only, a 2-D filter can be realized by the cascade of two 1-D filters in order to significantly reduce the hardware complexity for implementation. In many applications, this is sufficiently important to offset the slight degradation of the frequency response that occurs along the diagonal directions of the frequency plane [2.12], [2.22], [2.24], [2.26].

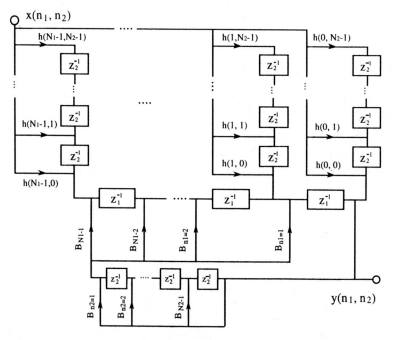

Fig. 2.10. Direct-form realization of a 2-D IIR filtering system with separable denominator.

Various design methods are available for 2-D IIR filters, whose implementation structures can vary significantly in terms of computational complexity, sensitivity to coefficient quantization, output noise level produced by the quantization of products and processing speed [2.2], [2.25], [2.28 − 2.30]. A possible example of a direct-form realization of a 2-D filter with separable denominator, i.e. described by the transfer function (2.11), is illustrated in Fig. 2.10.

Other forms of implementation are based on transformation methods, including transformations from analog filter models whereby a 2-D analog passive filter network is converted into a topologically related 2-D digital filter network, e.g. 2-D ladder structures [2.2], [2.31]. Although this preserves the fundamental properties of the original analog network, in particular lower sensitivity to coefficient quantization effects, it may lead to structures processing relatively lower speed of operation. Next, we shall look at various examples of 2-D filter structures actually implemented in integrated circuit form and compare their relative advantages from the viewpoints of performance and cost of implementation (basically area consumption) and operation (power dissipation).

2.4 HARDWARE IMPLEMENTATIONS OF 2-D FILTERS

2.4.1 General Considerations

2-D filters can be implemented using either digital or analog hardware. Digital 2-D filters possess several well-known advantages, such as highly efficient VLSI design methodologies and tools, flexible function programmability, high precision, low noise and distortion and high processing speed at the expense of power dissipation. Such features make digital 2-D implementation particularly suitable for large processing systems, as it is often needed in a variety of image processing applications [2.6 − 2.7], [2.12 − 2.13], [2.24], [2.30 − 2.35]. A general 2-D digital filtering implementation scheme is illustrated in Fig. 2.11, where A/D and D/A converters are required to convert the signals between the analog image and its digital processing core.

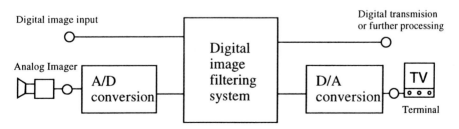

Fig. 2.11. A general 2-D digital filter implementation scheme.

Analog 2-D filters, on the other hand, can be implemented either in continuous-time or in discrete-time form. Continuous-time implementations can be based either in classical analog signal processing networks [2.9 – 2.11], [2.14 – 2.16] or in neural-based processing networks [2.8], [2.36]. Analog 2-D filters are particularly suitable for such applications as machine vision, robotics and portable equipment, where low power dissipation, small chip area, high processing speed and moderate processing accuracy are key factors. A general 2-D analog filtering implementation scheme is illustrated in Fig. 2.12, where analog image signals are directly processed instead of using the converters needed in digital implementations, and which dissipate more power, take more chip area and often introduce additional noise and distortion.

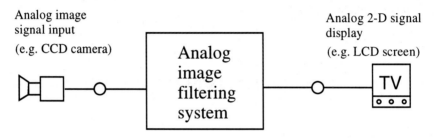

Fig. 2.12. A general 2-D analog filter implementation scheme.

Most image signal sources currently available are based on CCD and even CMOS imagers whose image output signals are already provided in sampled 2-D form. Moreover, many commonly used displays, notably LCD, are also directly driven by sampled 2-D analog signals. Consequently, discrete-time 2-D filter implementation techniques are particularly suitable for the general 2-D analog filter implementation scheme of Fig. 2.12. The design of such filters employs a variety of approaches well-developed for the digital 2-D domain and their integrated circuit implementations can produce more accurate and stable operations than their continuous-time counterparts.

Next, we describe some examples of 2-D filters actually implemented in integrated circuit form using both digital and SC techniques with on-chip delay-line (DL) memory blocks.

2.4.2 Examples of Digital VLSI 2-D Filter Implementations

A set of 1.2-µm CMOS chips for 2-D HDTV subband filtering with on-chip DL memories is described in [2.37]. Subband coding is a promising technique to reduce the high transmission rate of HDTV signals on digital channels. Basically, this consists of prefiltering and downsampling the image signals with a set of high- and low-pass filters in both the horizontal and vertical directions in order to generate four filtered subbands for further processing at lower rates. In each chip, a (10 x 14)th-order (vertical x horizontal) FIR filtering function is realized. Fig. 2.13 illustrates the chip outline where we can see the major implementing blocks. The relevant performance characteristics of the resulting prototype chips are summarized in Table 2.1.

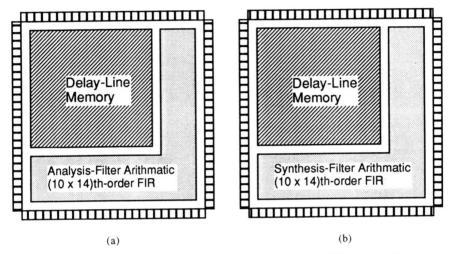

Fig. 2.13. Outline of the chip for the (a) 2-D analysis filter, and (b) 2-D synthesis filter.

Table 2.1
Performance characteristics of the 2-D filter chips in [2.37]

	Analysis Filter Chip	Synthesis Filter Chip
Active Area	92 mm^2	92 mm^2
Power Dissipation	600 mW	600 mW
Input Sampling Rate	72 MHz	18 MHz
Output Sampling Rate	18 MHz	72 MHz
Technology	1.2-μm CMOS	1.2-μm CMOS

Another example of a 54 MHz digital 2-D filter chip set for HDTV applications is reported in [2.38]. HDTV requires intensive image processing algorithms for real-time operations. In this chip set, a (16 × 16)th-order symmetric FIR filtering function together with a programmable DL memory block have been realized for implementing a bandwidth reduction algorithm. The outline of the filter chip and programmable DL memory chip are shown, respectively, in Fig. 2.14(a) and (b). The resulting relevant performance characteristics are given in Table 2.2 and Table 2.3, respectively for the filter chip and the programmable DL memory chip.

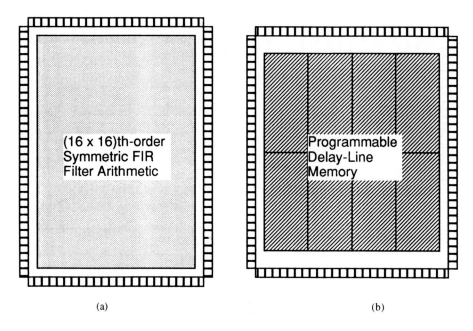

Fig. 2.14: Chip outline of (a) filter chip and (b) DL chip.

Table 2.2
Summary of the filter chip performance [2.38]

Active Area	35 mm^2
Clock frequency	60 MHz
Power Dissipation	500 mW @ 54 MHz
Technology	1.0-μm CMOS

Table 2.3
Summary of the programmable DL chip performance [2.38]

Active Area	3.2 x 3.8 mm^2
Clock frequency	80 MHz
Power Dissipation	100 mW @ 54 MHz

2.6.3 Examples of Analog VLSI 2-D Filter Implementations

A 1.2-μm CMOS analog video comb filter for luminance and chrominance separation of NTSC video signals is described in [2.39].

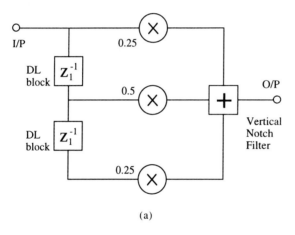

(a)

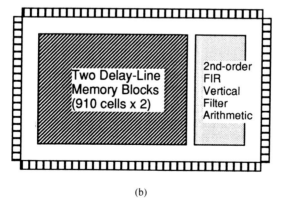

(b)

Fig. 2.15. (a) Block diagram and (b) chip outline of an analog video comb filter employing SC techniques for NTSC applications [2.39].

To achieve high image quality, 2-D filtering is employed for decoding the luminance and chrominance components from the composite video signal. For low cost implementation and low power dissipation, SC networks have been investigated to realize a solely vertical 2nd-order FIR filtering function with two DL blocks, each containing 910 storage cells. The simplified block diagram is illustrated in Fig. 2.15(a). The two largest blocks in the resulting chip, whose outline is shown in Fig. 2.15(b), correspond to the DL memory blocks. The resulting relevant performance characteristics are summarized in Table 2.4.

Table 2.4
Performance characteristics of the 2-D analog filter chip in [2.39]

Active area	11.7 mm^2
Power consumption	170 mW
Dynamic range	> 51 dB
Fixed pattern noise	-55 dB
Full-scale input voltage	2.6 V$_{p-p}$
Technology	1.2-µm CMOS

Another example of an analog 2-D filter employing SC techniques is reported in [2.10]. In this work, a (2 x 5)th-order SC FIR filtering system with two DL memory blocks has been implemented on a single chip, for both image highpass filtering and enhancement applications.

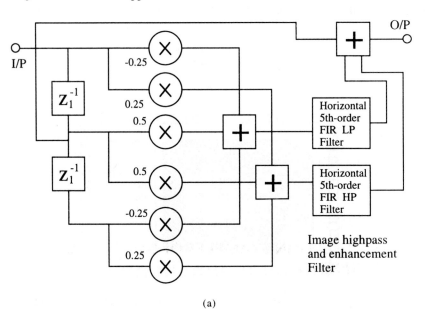

(a)

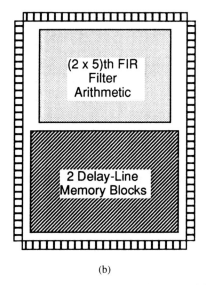

(b)

Fig. 2.16. (a) Block diagram and (b) chip outline of an analog 2-D highpass filter employing SC techniques [2.10].

The simplified block diagram is illustrated in Fig. 2.16(a). In the chip outline depicted in Fig. 2.16(b), the bottom half is occupied by the two large DL memory blocks whereas the top half includes 120 opamps together with the associated switches and capacitors. Table 2.5 summarizes the relevant chip performance.

Table 2.5
Performance characteristics of the 2-D analog filter chip in [2.10]

Chip size	5.4 x 6 mm^2
Clock rate	14.32 MHz
Single swing	2 $V_{p\text{-}p}$
Power dissipation	300 mW
Random noise	-66 dB
Technology	2.0-µm CMOS

2.5 APPLICATION EXAMPLES OF 2-D SIGNAL FILTERING

2-D signal processing has been used in many different areas, ranging from digital image processing for satellite photographs and radar maps, to medical x-ray images, video sequences, and even seismic signal processing. These are abundantly documented in a vast body of literature [2.1 – 2.2], [2.3], [2.6], [2.8], [2.40 – 2.41]. This section briefly outlines some of the application areas where the type of 2-D filtering functions and analog implementation techniques addressed in this book can be employed to achieve performance/cost advantages over digital VLSI implementations.

2.5.1 Noise Removal

Transmission noise in an image tends to be spatially independent and therefore its energy content is usually concentrated at frequencies higher than that of the image.

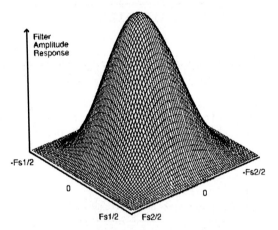

Fig. 2.17: Typical amplitude response of a 2-D lowpass filter system for noise remove.

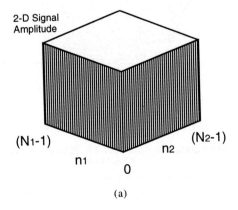

(a)

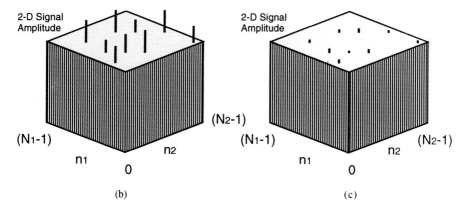

Fig. 2.18. (a) Ideal image in 3-D view; (b) image of (a) added by high-frequency noise; (c) resulting image after noise removal processing.

Consequently, by processing an image through a lowpass filtering function tends to remove a large amount of the noise content without changing the image significantly. A 2-D (3 x 3)rd-order FIR lowpass filter system is commonly used for such function typically yielding the 3-D amplitude response illustrated in Fig. 2.17. The application of such filter for the processing of a noise corrupted image signal can be observed in the sequence of images depicted in Fig. 2.18.

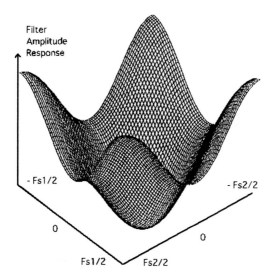

Fig. 2.19. Typical amplitude response of a 2-D highpass filtering system for edge enhancement.

2.5.2 Edge Enhancement

Many image-related applications such as pattern recognition and machine vision require images of objects with enhanced edges. In an optical image, the pixel intensity changes rapidly across edges, which means that the frequency content due to edges tends to be concentrated at the high end of the spectrum. As a consequence, edge enhancement can be achieved by using highpass filters. Fig. 2.19 illustrates the 3-D amplitude response of a 2-D (3 x 3) order FIR highpass filtering system.

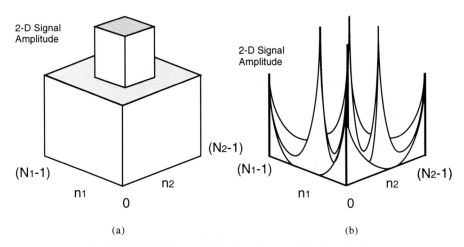

Fig. 2.20. (a) Ideal image. (b) Processed image with enhanced edges.

An application of such filter is depicted in Fig. 2.20, showing first the original ideal image, in Fig. 2.20(a), and then the resulting processed image with enhanced edges, in Fig. 2.20(b). Edge enhancement is often used in robotics vision systems where the mere determination of the geometric shape of an object to be handled by the robot is critical to the successful robot motion.

2.5.3 Applications in HDTV of 2-D Subband Filtering

Subband coding is a promising source coding scheme to reduce the high transmission data rate of HDTV signals over digital channels. Today, a HDTV codec with a transmission rate of about 140 Mbit/s can be implemented at acceptable hardware cost. The subband codec considered here is shown in Fig. 2.21. The HDTV input signal is split into four subbands containing different frequency domains of the input picture signal, as illustrated in Fig. 2.22. An input picture and the resulting four subbands are shown in Fig. 2.23. The 2-D lowpass filtered and downsampled subband I contains similar information as the input signal, but with a lower resolution. The 2-D highpass filtered subbands contain only information at the place of edges or corners in the input picture.

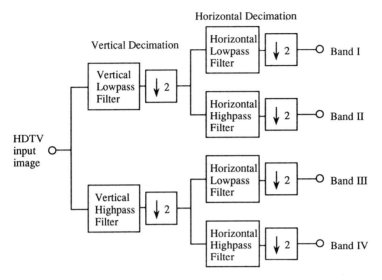

Fig. 2.21. Block diagram of a subband codec using a separable 2-D filter bank.

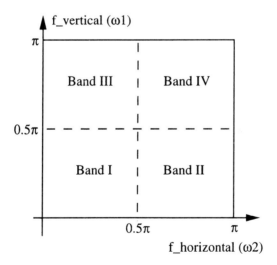

Fig. 2.22. Frequency band division into 4 subbands.

This information is important to maintain the sharpness of the HDTV picture. To allow visualization of the contents of the higher subbands, their contrast has been modified significantly.

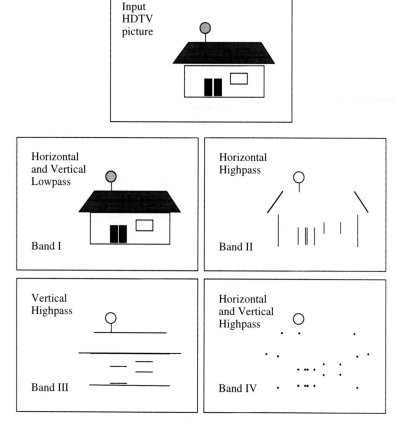

Fig. 2.23. Example of band separation using 2-D subband filtering.

As can be seen above, band II contains mainly horizontal lines, band III vertical lines and band IV diagonal lines or crossing points. The separation into these four subbands allows an efficient coding according to the properties of each subband by a motion compensated hybrid codec or quantizers followed by run length coders and variable length coders.

2.5.4 Applications of Video Image Special Effects

The use of image special effects, such as electronic zoom, image resizing which includes compression, expansion and rotation in television productions has experienced a rapid growth. By making the image size smaller, less data will be needed to transmit and store. This is the case, for example, when the video stream is resized to a small image size, e.g. CIF image size with 352 x 288 pixels, or QCIF sized with 176 x 144 pixels, sent it to its final destination and then zoomed back to a larger size for viewing. For a typical CCIR601 video image with the size of 720 x 240 pixels resized to a CIF sized video image, this leads to more than 40% reduction of the bandwidth needed for transmission. Such operation is implemented by appropriate 2-D filtering functions that prevent the occurrence of aliasing effects when the image signals are downsampled. Image resizing is an important function in a complete video compression system, such as the one illustrated by the block diagram of Fig. 2.24.

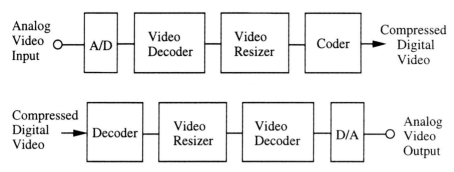

Fig. 2.24. Block diagram of a video compression system employing image resizing functions.

2.6 SUMMARY

This chapter introduced the basic concepts and properties of 2-D signals and filtering systems, both with continuous-time and discrete-time lines, described several examples of practical integrated circuit implementations, and outlined several application areas where the type of 2-D filtering functions and implementation techniques addressed in this book can be effectively employed. Given the rather large amount of data associated with 2-D signals, implementation efficiency is paramount for the successful realization of 2-D filtering systems in integrated circuit form. Illustrative examples were given of various digital and analog 2-D filtering chips where large portions of silicon area, both in absolute values and in proportion to the whole chip sizes, were dedicated to the implementation of the DL memory blocks that are essential for 2-D signal processing. This gives the fundamental motivation for the work described in the following chapters of this book, and which concerns the development of new multirate signal processing techniques that are particularly suitable to reduce the size of such DL memory blocks and hence lead to more cost effective implementations of 2-D filtering systems in integrated circuit form.

REFERENCES

[2.1] J. S. Lim and A. V. Oppenheim, *"Advanced Topics in Signal Processing"*, Prentice Hall, 1988.

[2.2] Wu-Sheng Lu and Andreas Antoniou, *"Two-Dimensional Digital Filters"*, Marcel Dekker, Inc., 1992.

[2.3] M.Soumekh, *"Fourier Array Imaging"*, Prentice Hall, 1994.

[2.4] V. Cappellini, A. G. Constantinides and P, Emiliami, *"Digital Filters and their Applications"*, Academic Press, 1978.

[2.5] R.C. Gonzalez and Paul Wintz, *"Digital Image Processing"*, Addison-Wesley Publishing Company, 1977.

[2.6] B. Ackland, "VLSI Architectures for Multimedia and Video Conferencing", *Proc. IEEE International Symposium on Circuits and Systems*, London, pp. 147, Chapter 3.21, Tutorial, 1994.

[2.7] T. Fujio, "HDTV Systems", *Proc. of the IEEE*, Vol. 73, No. 4, pp. 646-655, April, 1985.

[2.8] C. Mead, *"Analog VLSI and Neural Systems"*, Addison-Wesley Publishing Inc., 1989.

[2.9] K. Nishikawa, T. Takebe and M. Hayashihara, "Two-Dimensional Switched-Capacitor Filter.", *IEEE International Symposium on Circuits and Systems*, pp. 73-76, Montreal, Canada, June 1984.

[2.10] K.Matsui, T.Matsuura, S.Fukasawa, Y.Izawa, Y.Toba, N.Miyake and K.Nagasawa, "CMOS Video Filters Using SC 14-MHz Circuits", *IEEE Journal of Solid-State Circuits*, Vol. sc-20, No. 6, pp. 1096-1101, December 1985.

[2.11] M. A. Sid-Ahmed, "Two-Dimensional Analog Filters: A New Form of Realization", *IEEE Transactions on Circuits and Systems*, vol. 36, no. 1, pp. 153-154, January 1989.

[2.12] M. L. Liou and T. Russell, "An Overview for Video Signal Processing", *Proc. IEEE International Symposium on Circuits and Systems*, pp. 208-211, PA, May, 1987.

[2.13] Y.Yasumoto, H. Sakashita, N. Yamaguchi and K. Yamamoto, "Digital Video NTSC & PAL Signal Processor for VLSI", *Proc. IEEE International Symposium on Circuits and Systems*, pp. 172-175, PA, May, 1983.

[2.14] H. J. Kaufman and M. A. Sid-Ahmed, "2-D Analog Filters for Real -Time Video Signal Processing", *IEEE Transactions on Consumer Electronics*, vol. 36, no.2, pp. 137-140, May 1990.

[2.15] M. A. Sid-Ahmed, "Realization of 2-D IIR Filters Using Sampled-and-Hold Circuitry", *IEEE International Symposium on Circuits and Systems*, pp. 2467-2470, June 1991.

[2.16] A. Handkiewicz, "Two-Dimensional SC Filter Design Using a Gyrator-Capacitor Prototype", *International J. of Circuit Theory and Applications*, Vol. 16, pp. 101-105, 1988.

[2.17]   A. Handkiewicz, "Two-Dimensional Switched-Capacitor Filter Design System for Real-Time Image Processing", *IEEE Transactions on Circuits and Systems for Video Technology*, vol. 1, no.3, pp. 241-246, September 1991.

[2.18]   Alan V. Oppenheim, *"Digital Signal Processing"*, Prentice-Hall, Englewood Cliffs, NJ, 1975.

[2.19]   R.A. Gabel and R.A. Roberts, *"Signals and Linear Systems"*, John Wiley & Sons, Inc., 1980.

[2.20]   Athanasios Papoulis, *"Circuits and Systems - A Modern Apprroach"*, Holt, Rinehart and Winston, Inc., 1980.

[2.21]   R. Ansari and C.L. Lau, "2-D IIR Filters for Exact Reconstruction in Tree-Structured Sub-Band Decomposition", *Electronics Letters*, vol. 23, No.12, pp. 633-634, 4th June 1987.

[2.22]   Q.S. Gu, M.N.S. Swamy, L.C.K. Lee and M.O. Ahmad, "IIR Digital Filters for Sampling Structure Conversion and Deinterlacing of Video Signals", *IEEE International Symposium on Circuits and Systems*, pp. 973-976, San Diago, May 1995.

[2.23]   G. Schamel, "Pre- and Post-filtering of HDTV Signals for Sampling Rate Reduction and Display Up-conversion", *IEEE Transactions on Circuits and Systems*, vol. 34, no.11, pp. 1432-1439, November 1987.

[2.24]   U. Pestel and K. Gruger, "Design of HDTV Subband Filterbanks Considering VLSI Implementation Constraints", *IEEE Transactions on Circuits and Systems for Video Technology*, vol. 1, no.1, pp. 14-21, March 1991.

[2.25]   A.N. Venetsanopoulos and B.G. Mertzios, "A Decomposition Theorem and Its Implications to the Design and Realization of 2-D Filters", *IEEE Transactions on Acoustics, Speech, and Signal Processing*, vol. 33, no. 6, pp. 1562-1574, December 1985.

[2.26]   G.X. Gu and B.A. Shenoi, "A Novel Approach to the Synthesis of Recursive Digital Filters with Linear Phase", *IEEE Transactions on Circuits and Systems*, vol. 38, no. 6, pp. 602-612, June 1991.

[2.27]   H. Jaggernauth and A.N. Venetsanopoulos, "Real-Time Image Processing Through Distributed Arithmetic", *IEEE International Symposium on Circuits and Systems*, pp. 394-397, June 1983.

[2.28]   J. Jaggernauth, A.C.P. Loui and A.N. Venetsanopoulos, "Real-Time Image Processing by Distributed Arithmetic Implementation of 2-D Digital Filters", *IEEE Transactions on Acoustics, Speech, and Signal Processing*, vol. 33, no. 6, pp. 1546-1555, December 1985.

[2.29]   R. Gnanasekaran, "2-D Filter Implementations for Real-Time Signal Processing", *IEEE Transactions on Circuits and Systems*, vol. 35, no. 5, pp. 587-590, May 1988.

[2.30]   D. Raghuramireddy, X. Nie and R. Unbehauen, "Minimal and Low-Sensitivity Implementations of a class of 2-D Digital Filters", *IEEE International Symposium on Circuits and Systems*, pp. 718-721, June 1992.

[2.31]   C.J. Kulach, L.T. Bruton and N.R. Bartley, "A Real-Time Video Implementation of A 3-D 1st-Order Recursive Discrete-Time Filter", *IEEE International Symposium on Circuits and Systems*, pp. 609-612, Atlanta, GA., May 1996.

[2.32] C. Joanblanq, P. Senn and M.J. Colaitis, "A 54 MHz CMOS Programmable Video Signal Processor for HDTV Applications", *IEEE Journal of Solid-State Circuits,* Vol. sc-25, No. 3, pp. 730-734, June 1990.

[2.33] W.T. Mayweather, "A Video Rate Rerastering IC", *IEEE Custom Integrated Circuits Conference,* pp. 2521-2524, June 1990.

[2.34] K.Aono, M. Maruyama, T. Mori, H. Yamada and K. Hataya, "Implementation of a Bipolar Real-Time Image Signal Processor - RISP-II", *IEEE Journal of Solid-State Circuits,* vol. 22, no.3, pp. 403-408, June 1987.

[2.35] K. Ramachandran and R.R. Cordell, "A 30-MHz Programmable CMOS Video FIR Filter and Correlator", *IEEE International Symposium on Circuits and Systems,* pp. 705-708, June 1988.

[2.36] P.Kinget and M.S.J.Steyaert, "A Programmable Analog Cellular Neural Network CMOS Chip for high speed Image Processing", *IEEE Journal of Solid-State Circuits,* vol. 30, no.3, pp. 235-243, March 1995.

[2.37] M. Winzker, K. Gruger, W. Gehrke and P. Pirsch, "VLSI Chip Set for 2-D HDTV Subband Filtering with On-Chip Line Memories", *IEEE Journal of Solid-State Circuits,* Vol. sc-28, No. 12, pp. 1354-1361, December 1993.

[2.38] C. Joanblanq, F. Rothan and P. Senn, "A 54 MHz Chip Set for HDTV Filtering", *IEEE Custom Integrated Circuits Conference,* pp. 2531-2534, June 1990.

[2.39] K. A. Nishimura and P. R. Gray, "A Monolithic Analog Video Comb Filter in 1.2-μm CMOS", *IEEE J. of Solid-State Circuits,* Vol. 28, No. 12, pp. 1331-1339, December 1993.

[2.40] P.P. Vaidyanathan, *"Multirate Systems and Filter Banks",* Prentice-Hall, Englewood Cliffs, NJ, 1993.

[2.41] C. Ngo, "Image Resizing and Enhanced Digital Video Compression", *EDN* , pp. 145-155, January 1996.

3

FUNDAMENTAL ASPECTS OF 2-D DECIMATION FILTERS

3.1 INTRODUCTION

In the previous chapter we reviewed basic concepts, forms of implementation and some applications of non-multirate two-dimensional (2-D) filters, both analog (continuous-time) and discrete-time. In both cases, the need to store complete line segments represents the major difficulty for implementation, given the required large size of the delay-line (DL) memory blocks and their impact on area and power consumption. This might be particularly disadvantageous when targeting low cost implementations, mostly employing switched-capacitor (SC) techniques.

In this chapter we discuss 2-D multirate signal processing techniques, particularly decimation, with the purpose of reducing the required size of the DL memory blocks in 2-D filters. For clarity of explanation, we begin in Section 3.2 with a brief review of fundamental aspects of decimation filters for the simpler case of one-dimensional (1-D) signals. In Section 3.3, these are also extended to the case of 2-D discrete-time signals. Then, Section 3.4 discusses the size requirements of the key DL memory blocks needed for realizing 2-D filters, both in the more traditional (non-multirate) form of implementation and in the new decimation form of implementation described in this book. A summary of the chapter is given in Section 3.5.

3.2 1-D DECIMATION

3.2.1 Basic Concepts

The emergence of 1-D digital multirate filtering [3.1 – 3.4] was motivated by the need to reduce the processing rate of many digital filtering applications, specially those dealing with signal frequencies close to the processing speed capabilities of the available technology. There are two basic forms of multirate processing, namely interpolation and decimation. Interpolation is concerned with the upsampling of a signal to increase its useful bandwidth, whereas decimation is the complementary process concerned with the downsampling of a signal to reduce unused bandwidth.

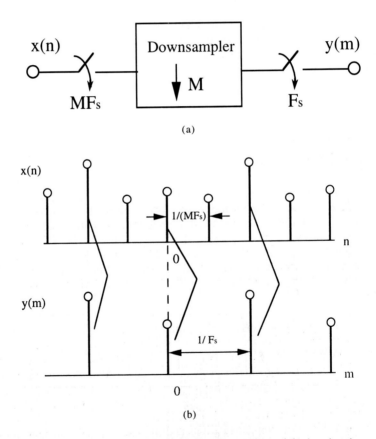

Fig. 3.1. M-fold downsampler. (a) Symbolic representation and (b) time-domain interpretation for $M = 2$.

By adequately using either one or both multirate processes combined together, it is always possible to adapt the sampling frequency to the bandwidth of the signal at each stage of a processing chain in order to best match the computation capabilities of the implementing hardware. Next, we shall review the basic definitions and properties of *decimation* which, in the context of this book, is particularly important to attack the problem of reducing the size of the DL memory blocks needed for the realization 2-D filtering functions.

Consider an M-fold downsampler, as shown in Fig. 3.1, which takes an input sequence $x(n)$ sampled at MF_s and produces an output sequence sampled at the lower sampling frequency F_s, and which can be described mathematically by

$$(3.1) \ldots \quad y(m) = x(Mm).$$

The downsampling factor M can, in general, be any positive real number, although in the context of this book only positive integer factors will be considered.

In the frequency domain, the operation of downsampling is illustrated in Fig. 3.2. Due to the discrete-time nature of the signals, the input signal spectrum is periodic with the input sampling frequency MF_s whereas the output signal spectrum is periodic with the lower output sampling frequency F_s.

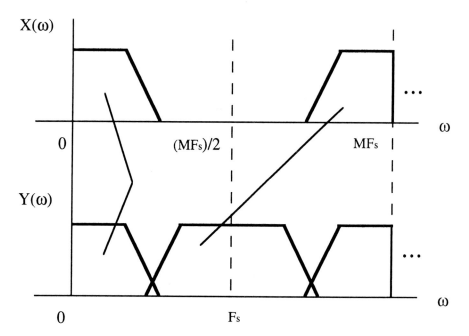

Fig. 3.2. The downsampling of a discrete-time signal translates the signal to lower frequencies.

Because of the frequency translation to lower frequencies, the downsampler should be in general preceded by a lowpass filter, the *decimation filter* shown in Fig. 3.3(a), which ensures that the signal being decimated is adequately bandlimited to prevent the occurrence of aliasing distortion. Usually, the decimation filter and downsampler are combined together in a single processing block designated by *decimator*. In order to take advantage of the reduction of the sampling frequency, it is possible to design such decimator for implementation in dedicated architectures where the operating speed of the active components is carried-out at the lower output sampling frequency, as conceptually indicated in Fig. 3.3(b). These are the polyphase architectures discussed next.

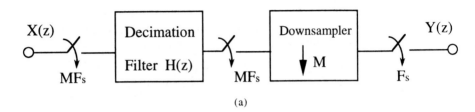

(a)

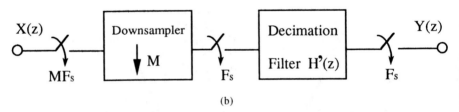

(b)

Fig. 3.3: Representation of (a) the general block diagram of a decimator and (b) its efficient form of implementation to take advantage of the reduced output sampling frequency.

3.2.2 Polyphase Implementation

Consider a prototype FIR filter represented by the z-transfer function

$$(3.2) \quad H(z) = \sum_{n=0}^{N-1} h(n) z^{-n},$$

where the unit delay z^{-1} corresponds to the input sampling frequency MF_s and N is the length of prototype filter. For simplicity of explanation we assume that $N =$

ML, where M is the decimation factor and L an integer greater than one. In such case, the z-transfer function can also be expressed as

$$(3.3) \quad H(z) = \sum_{m=0}^{M-1} G_m(z) z^{-m} ,$$

corresponding to the polyphase representation of the above filter (3.2), and such that the z-transfer functions of the polyphase sub-filters are given by

$$(3.4) \quad G_m(z) = \sum_{i=0}^{L-1} h(m+iM) z^{-iM} .$$

From (3.3) and (3.4) we obtain the direct-form polyphase decimation filter structure shown in Fig. 3.4 [3.1], with M polyphase sub-filters. Here, we can observe that while the overall structure maintains the high input sampling frequency MF_s, the coefficient multiplication within the various sub-filters are carried-out at the lower output sampling frequency F_s and hence taking advantage of the frequency downsampling from the input to the output of the decimator. As mentioned before, this is particularly relevant for reducing the speed and power requirements of the implementing circuits, specially when targeting very high-frequency applications [3.5 – 3.7].

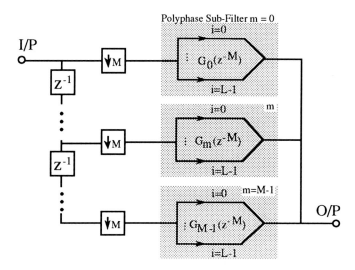

Fig. 3.4. 1-D polyphase structure for an FIR decimation filter with M-fold sampling rate reduction.

Consider now a prototype decimation IIR filter represented by the z-transfer function

$$(3.5)\ldots\quad H(z) = \frac{N(z)}{D(z)} = \frac{\sum_{i=0}^{N-1} a_i z^{-i}}{1 + \sum_{i=1}^{D} b_i z^{-i}} = \frac{a_0 + a_1 z^{-1} + \ldots + a_{N-1} z^{-(N-1)}}{1 + b_1 z^{-1} + \ldots + b_{N-1} z^{-(N-1)}},$$

where, for simplicity of explanation, it is assumed that the orders of the numerator and denominator are equal, i.e. $D = N - 1$. For a given M-fold decimation, it is possible to modify the above transfer function such that the delay terms in the denominator polynomial are all function of integer powers of z^{-1}. This is obtained by applying the well known multirate transformation [3.1 – 3.2] yielding

$$(3.6)\ldots\quad H(z) = \frac{N(z)}{D(z)} = \frac{\sum_{m=0}^{M-1} A_m(z^M) \cdot z^{-m}}{1 + \sum_{n=1}^{N-1} B_n(z^M)^{-n}}.$$

In the above expression, the coefficients of the numerator polynomial function are given by

$$(3.7)\ldots\quad A_m(z^M) = \sum_{n=0}^{N-1} a_{m+nM} z^{-nM},$$

where the modified coefficients a_{m+nM} and B_n, respectively in (3.7) and (3.6), are determined from the multirate transformation [3.1 – 3.2]. From expressions (3.6) and (3.7) we obtain the polyphase structure illustrated in Fig. 3.5, combining the type of direct-form polyphase structure given in Fig. 3.4 together with a classical direct-form recursive structure.

In this case, not only the coefficient multiplications within the polyphase sub-filters are carried-out at the lower output sampling frequency, as before, but also the recursive structure is also fully computed at the lower output sampling frequency. Most of today's efficient implementations of decimation filters, both FIR and IIR, are based on the 1-D polyphase structures described above. In general, such polyphase structures have the advantage that the polyphase sub-filters can be easily realized using efficient discrete-time signal processing techniques such as fast convolution methods based on the FFT [3.1 – 3.4]. Besides, in the direct-form of implementation it is also possible to exploit the symmetry of some impulse responses to obtain an additional reduction in computation by a factor of approximately two.

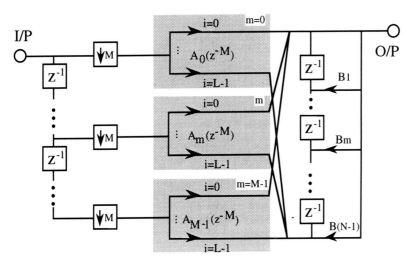

Fig. 3.5. 1-D polyphase structure for an IIR decimation filter with M-fold sampling rate reduction.

Next, we shall discuss the extension of the above multirate decimation techniques also to the case of 2-D discrete-time signals that are of particular relevance to the work in this book.

3.3   2-D DECIMATION

3.3.1   Basic Concepts

In modern applications of image processing, original images are often re-sized to a smaller image-size in order to allow such operations as video-in-a-window [3.5], [3.8 – 3.10]. Changing the size of an image, i.e. changing the number of pixels that are normally used to represent the image, also involves changing the sampling frequencies of the original image. For example, to shrink an image without sacrificing too much of the signal image information it is necessary to limit the bandwidth of the signal to comply with the Nyquist sampling theorem. To accomplish this correctly it is required the use of 2-D decimation filters.

Consider a 2-D (M_1, M_2)-fold downsampler, as shown in Fig. 3.6, which takes an image sequence $x(n_2, n_1)$ with sampling rates $M_1 F_{s1}$ and $M_2 F_{s2}$, and produces the output image sequence expressed as

(3.8) ... $y(m_2, m_1) = x(M_2 m_2, M_1 m_1),$

with lower sampling rates F_{s1} and F_{s2}. In the above expression, M_1, M_2 are positive integer numbers called, respectively, the vertical and horizontal decimation factors.

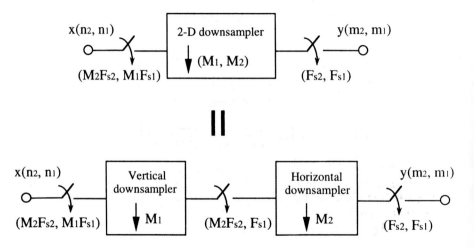

Fig. 3.6. 2-D downsampler with decimation ratio (M_1, M_2).

The time-domain interpretation of the operation of such 2-D downsampler is illustrated in Fig. 3.7. For the original discrete-time 2-D image signal depicted in Fig. 3.7(a), the horizontal decimation illustrated in Fig. 3.7(b) produces only one output pixel for every M_2 input pixels in each line and thus reduces by M_2 the total number of output pixels needed to be stored in the DL memory blocks. In the frequency domain, this is equivalent to compressing by M_2 and frequency-translating the spectrum of the horizontal signal, as schematically illustrated in Fig. 3.7(c) referring to the input spectrum as previously shown in Fig. 2.4(b). Because of the potential for aliasing distortion caused by such spectral compression and translation, the appropriate decimation filter must precede the operation of downsampling, as in the case of 1-D decimation.

Multirate Switched-Capacitor Circuits for 2-D Signal Processing 47

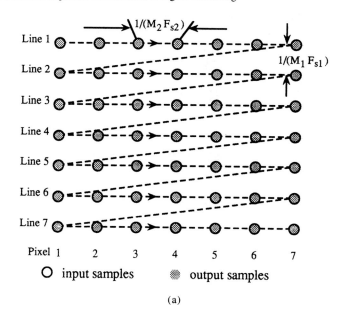

(a)

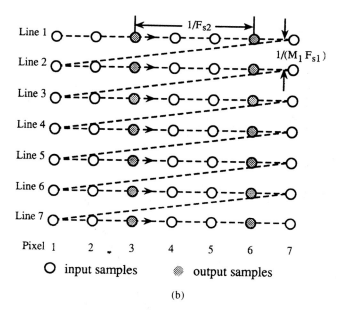

(b)

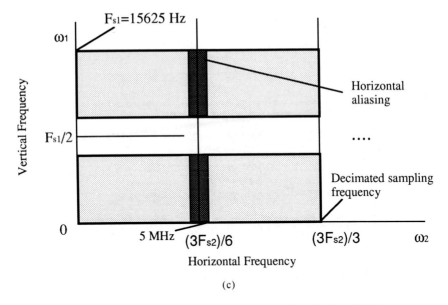

(c)

Fig. 3.7: (a) Original discrete-time 2-D input image signal. (b) Decimated 2-D image, by $M_2 = 3$, in the horizontal dimension. (c) Resulting compressed frequency spectrum.

The above concepts can be easily extended also to the vertical dimension of an image such that, as schematically illustrated in Fig. 3.8(a), only one output line for every M_1 input lines is produced in a given image frame.

(a)

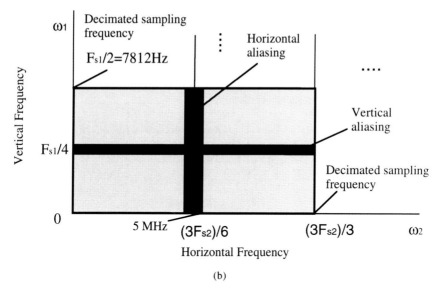

Fig. 3.8. (a) Decimation occurring in both dimensions, with $M_1 = 2$, $M_2 = 3$.
(b) Resulting frequency spectrum.

Moreover, when combining together the horizontal and vertical decimation operations the total number of pixels in a given image frame, and hence the required memory size for processing, is reduced by a factor $M_1 M_2$ of the combined horizontal and vertical decimation factors. The resulting frequency compression in both dimensions can also be observed in the frequency domain representation illustrated in Fig. 3.8(b).

Besides the already mentioned reduction of the size of the DL memory blocks, the above discussion suggests that, similarly to the 1-D case, the 2-D decimation can also have a very big impact in the reduction of the speed requirements [3.4 – 3.5], [3.11 – 3.12]. This, however, can only be achieved if the corresponding implementations are based on the 2-D polyphase structures discussed next.

3.3.2 Polyphase Implementation

Similarly to the 1-D case discussed in Section 3.2, a 2-D decimation filter can be implemented either in the non-efficient form illustrated in Fig. 3.9(a) or in the more efficient form represented in Fig. 3.9(b). It is this latter form of implementation, where processing is effectively carried-out at the lower sampling frequencies (F_{s1}, F_{s2}) rather than at the higher sampling frequencies ($M_1 F_{s1}$, $M_2 F_{s2}$), that permits to take full advantage of the process of downsampling to reduce the speed requirements of the active components. However, since this form of implementation may require the factorization of 2-D polynomials, and which has no general solution for the efficient implementation of 2-D IIR z-transfer functions [3.4], it is necessary to constraint its applicability to those 2-D transfer functions that would not lead to unrealistic complexity for hardware implementation. For example, it can easily be shown that transforming a high-order non-separable 2-D IIR decimation filter would cause the resulting filter to become too complex for practical realization [3.4], [3.13 – 3.14]

In the following derivation of the polyphase structure for the efficient implementation of 2-D FIR decimators we will use, for clarity of explanation, an example with a 2-fold decimation in both dimensions, i.e. $(M_1, M_2) = (2, 2)$. In this case, a general decimation filter $H(z_2, z_1)$ can be decomposed as

(3.9) ...

$$H(z_1, z_2) = H_1(z_1, z_2) + z_2^{-1} H_2(z_1, z_2) + z_1^{-1} H_3(z_1, z_2) + z_2^{-1} z_1^{-1} H_4(z_1, z_2)$$

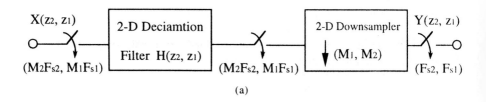

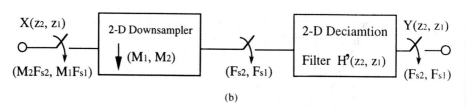

Fig. 3.9. (a) A general implementation of the 2-D decimation filter and (b) the efficient implementation of such system.

leading to the block diagram representation shown in Fig. 3.10, where the 2-D z-transfer functions of sub-filters H_1, H_2, H_3 and H_4 are directly decomposed from the original 2-D prototype filter [3.4].

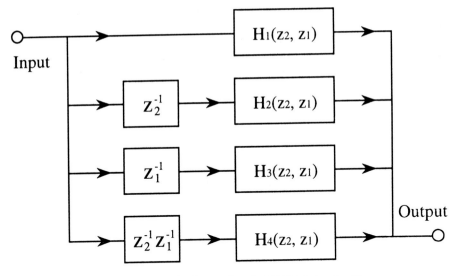

Fig. 3.10. Polyphase implementation of a 2-D FIR decimation filter.

When the decimation filter function $H(z_2, z_1)$ is separable, i.e. it can be expressed as

(3.10) ... $\quad H(z_2, z_1) = H_1(z_1)H_1(z_2),$

then an efficient implementation can be obtained by considering separate 1-D decimation operations in both dimensions. This yields

(3.11) ... $\quad H(z_1, z_2) = (\sum_{m_1=0}^{M_1-1} H_{1m_1}(z_1) z_1^{-m_1})(\sum_{m_2=0}^{M_2-1} H_{2m_2}(z_2) z_2^{-m_2}),$

where $H_{1m1}(z_1)$ and $H_{2m2}(z_2)$ are, respectively, the vertical and horizontal polyphase sub-filters with either FIR or IIR z-transfer functions. Fig. 3.11 shows the case of an FIR filter where the input sampling frequency is the same as in the previous structure of Fig. 3.10, but where the polyphase sub-filters have now been placed after downsampling and hence are allowed to operate at the lower output sampling frequencies.

Having established the efficient polyphase structures for implementation of 2-D decimation filters, specially in separable form, we now proceed to investigate the resulting size requirements for the DL memory blocks needed for implementation and discuss the comparative advantages with respect to the equivalent non-efficient forms of implementation.

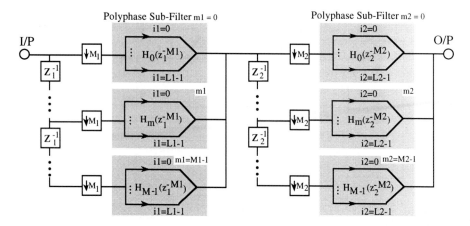

Fig. 3.11. Efficient polyphase implementation for a 2-D separable decimation filter.

3.4 DL MEMORY BLOCK REQUIREMENTS FOR 2-D FILTERS

3.4.1 Full-Size DL for 2-D Non-Multirate Filters

When filtering a 2-D signal, the current output *pixel* (sampled signal point) is generally a function of previous pixels that are not only from the current line but also from previous lines. Hence, in order to retain the pixel information from previous line-segments it is mandatory to employ DL memory blocks. Since the number of pixels per line is usually high, the required storage capacity of each DL memory block is also high and hence the size of the implementing circuit is large. This is a major difficulty for the implementation of analog 2-D filters, specially when targeting low cost solutions to compete with the dominant digital VLSI forms of implementation. The purpose of the 2-D decimation techniques introduced before is to allow a substantial reduction of the size of the DL memory blocks by adapting more efficiently the input/output sampling of the signals to the useful bandwidth for processing.

As illustrated in Fig. 3.12, the unit delay element (z_1^{-1}) in the vertical dimension of a 2-D image signal represents a DL memory block containing a cascade of N unit-delays z_2^{-1} of the sampling frequency in the horizontal dimension.

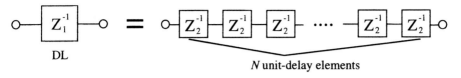

Fig. 3.12. Symbolic representation of a DL memory block.

The length N of the DL memory block is determined by the ratio of the horizontal and vertical sampling frequencies, i.e.

(3.12) ...   DL_length = F_{s2} / F_{s1},

and therefore varies for different image sampling frequencies (F_{s1}, F_{s2}). Consider, for example, a PAL video image where the horizontal and vertical sampling frequencies, respectively, are usually $F_{s2} = 17.718$ MHz and $F_{s1} = 15.625$ kHz. From (3.12) this leads to a total of 1140 storage cells (z_2^{-1} unit-delays) needed in the DL memory blocks to memorize a full line-segment signal.

Since the realization of high order 2-D filters requires that several of such large DL memory blocks are employed for vertical filtering, the DL memory blocks alone are responsible for most of the silicon area taken up in an integrated circuit implementation. Besides area, large analog DL memory blocks require high driving capability of the active components because of the capacitive nature of the storage cells [3.15 – 3.20]. This, in turn, also contributes to add noise and increase distortion. Next, we shall see how 2-D decimation techniques can be efficiently used to reduce the size of DL memory blocks and hence have the potential to lead to more compact integrated circuit implementations of 2-D filtering functions, with lower power requirements and also more immune to performance degradation due to noise and distortion effects.

3.4.2   Reduced DL Size for 2-D Decimation Filters

Consider first the block diagram of Fig. 3.13(a) where the 1-D lowpass filtering system $H(z)$, with bandlimited input signal, is followed by a delay chain of 10 unit-delays. An equivalent multirate system is given by the block diagram of Fig. 3.13(b), where a 5-fold downsampler followed by a 5-fold upsampler and an ideal anti-image filter are inserted between those unit-delays and the output terminal. The downsampler in Fig. 3.13(b) can now be placed in front of the delay chain such that only two 5-unit delays are employed after the filter $H(z)$, as illustrated in Fig. 3.13(c). By moving the downsampler further to the front of filter, as shown in

Fig. 3.13(d), $H(z)$ becomes $H'(z^5)$ and hence the operating speed of the system is also reduced 5-fold since the unit delay z^{-1} is now $(z^5)^{-1}$.

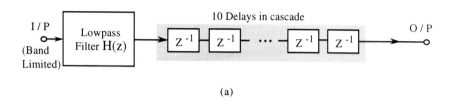

(a)

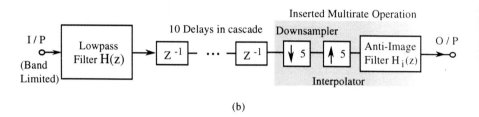

(b)

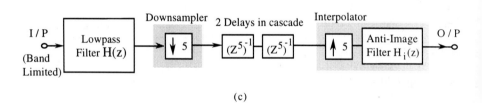

(c)

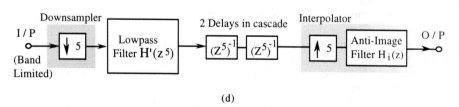

(d)

Fig. 3.13: (a) Traditional 1-D filtering system cascaded with a delay chain of 10 unit-delays. (b) Equivalent multirate operation. (c) Moving the downsampler to the front of the delay chain. (d) Moving the downsampler to the front of the decimation filter.

Now, consider that the system of Fig. 3.13(a) is a 2-D filter where $H(z)$ is the horizontal filter and the delay chain represents a simple vertical filter with DL length of 10 memory cells. From the multirate system of Fig. 3.13(d) we see that the length of the DL memory block is also reduced 5-fold. The 5-fold downsampler together with the associated filtering function $H'(z^5)$ is defined as the decimation filter in the horizontal dimension and which, besides the required baseband filtering, should also provide the necessary anti-aliasing filtering for the incoming out-of-band signals.

In general, an M_2-fold horizontal decimation filter reduces by a factor of M_2 the total number of pixels stored for each line segment. If a similar operation is also applied in the vertical dimension, using an M_1-fold vertical decimation filter, then the total capacity needed for the complete frame-memory will be reduced by a factor of $M_1 M_2$. This can lead to very significant savings in chip area and power dissipation and hence render the analog implementation of some specialized 2-D filtering applications rather competitive with their digital counterparts.

3.5 SUMMARY

This chapter discussed the use of 2-D multirate signal processing techniques, particularly decimation, with the purpose of reducing the size of the DL memory blocks that are needed for the integrated circuit implementation of 2-D filters. This is mostly based on the extension to 2-D signals of the basic concepts and implementing structures already known for 1-D signals. However, unlike in the case of 1-D filtering functions that can readily be implemented based on classical polyphase structures, practical considerations led us to constraint the implementation of 2-D decimation filters to the case of separable z-transfer functions, where the horizontal and vertical decimation functions are treated separately as 1-D filters.

The use of such decimation techniques can lead to a reduction of the size requirements of the DL memory blocks by a factor equal to the horizontal decimation factor. When, besides the horizontal dimension, decimation is also applied to the vertical dimension then the reduction of the frame-memory size is proportional to the product of both horizontal and vertical decimation factors. Overall, such reductions of memory size requirements for 2-D signal processing have a great impact in the reduction of the total area needed for integrated circuit implementation, as well as in the reduction of power dissipation and non-ideal noise and distortion effects.

REFERENCES

[3.1] R. Crochiere and L.R. Rabiner, *"Multirate Digital Signal Processing"*, Prentice-Hall, Englewood Cliffs, NJ, 1983.

[3.2] Maurice Bellanger, *"Digital Processing of Signals"*, Academic Press, 1978.

[3.3] N.J. Fliege, *"Multirate Digital Signal Processing"*, John Wiley & Sons, Inc., 1994.

[3.4] P.P. Vaidyanathan, *"Multirate Systems and Filter Banks"*, Prentice-Hall, Englewood Cliffs, NJ, 1993.

[3.5] U. Pestel and K. Gruger, "Design of HDTV Subband Filterbanks Considering VLSI Implementation Constraints", *IEEE Transactions on Circuits and Systems for Video Technology*, vol. 1, no.1, pp. 14-21, March 1991.

[3.6] G. Schamel, "Pre- and Post-filtering of HDTV Signals for Sampling Rate Reduction and Display Up-conversion", *IEEE Transactions on Circuits and Systems*, vol. 34, no.11, pp. 1432-1439, November 1987.

[3.7] R.P. Martins and J.E. Franca, "A 2.4-μm CMOS SC Video Decimator with Sampling Rate Reduction from 40.5 MHz to 13.5 MHz", *IEEE Custom Integrated Circuits Conference*, pp. 1-4, San Diego, May 1989.

[3.8] M. Harrand, M. Henry, P. Chaisemartin, P Mougeat, Y. Durand, A. Toumier, R. Wilson, J. Herluison, J. Longchambon, J. Bauer, M. Runtz and J. Bulone, "A Single Chip Videophone Video Encode/Decoder", *Proc. 1995 IEEE International Solid-State Circuits Conference*, pp. 292-293.

[3.9] B.M. Gordon, T.H. Meng and N. Chaddha, "A 1.2 mW Video-Rate 2-D Color Subband Decoder", *Proc. 1995 IEEE International Solid-State Circuits Conference*, pp. 290-291.

[3.10] C. Ngo, "Image Resizing and Enhanced Digital Video Compression", *EDN*, pp. 145-155, January 1996.

[3.11] T. C. Chen and R. J. P. De Figueiredo, "Image Decimation and Interpolation Based on Frequency Domain Analysis", *IEEE Transactions on Communications*, Vol. com-32, No. 4, pp. 479-484, April 1984.

[3.12] M. L. Liou and T. Russell, "An Overview for Video Signal Processing", *Proc. IEEE International Symposium on Circuits and Systems*, pp. 208-211, PA, May, 1987.

[3.13] R. Ansari and C.L. Lau, "2-D IIR Filters for Exact Reconstruction in Tree-Structured Sub-Band Decomposition", *Electronics Letters*, vol. 23, No.12, pp. 633-634, 4th June 1987.

[3.14] Q.S. Gu, M.N.S. Swamy, L.C.K. Lee and M.O. Ahmad, "IIR Digital Filters for Sampling Structure Conversion and Deinterlacing of Video Signals", *IEEE International Symposium on Circuits and Systems*, pp. 973-976, San Diago, May 1995.

[3.15] K.Matsui, T.Matsuura, S.Fukasawa, Y.Izawa, Y.Toba, N.Miyake and K.Nagasawa, "CMOS Video Filters Using SC 14-MHz Circuits", *IEEE Journal of Solid-State Circuits*, Vol. sc-20, No. 6, pp. 1096-1101, December 1985.

[3.16]   H. J. Kaufman and M. A. Sid-Ahmed, "2-D Analog Filters for Real-Time Video Signal Processing", *IEEE Transactions on Consumer Electronics*, vol. 36, no.2, pp. 137-140, May 1990.

[3.17]   K. A. Nishimura and P. R. Gray, "A Monolithic Analog Video Comb Filter in 1.2-µm CMOS", *IEEE J. of Solid-State Circuits*, Vol. 28, No. 12, pp. 1331-1339, December 1993.

[3.18]   D. Gerna, M. Brattoli, E. Chioffi, G. Colli, M. Pasotti and A. Tomasini, "An Analog Memory for a QCIF Format Image Frame Storage", *IEEE International Symposium on Circuits and Systems*, pp. 289-292, Atlanta, May 1996.

[3.19]   C.A. Mead and T. Delbruck, "Scanners for Visualizing Activity of Analog VLSI Circuit", *Analog Integrated Circuits and Signal Processing*, No. 1, pp. 93-106, June 1991.

[3.20]   E. Franchi, M. Tartagni, R. Guerrieri and G. Baccarani, "Random Access Analog Memory for Early Vision", *IEEE J. of Solid-State Circuits*, Vol. 27, No. 7, pp. 1105-1109, July 1992.

4

POLYPHASE-COEFFICIENT STRUCTURES FOR 2-D DECIMATION FILTERS

4.1 INTRODUCTION

As mentioned in previous chapters, the implementation of 2-D SC filtering functions using multirate processing techniques is motivated by the need to reduce both the processing speed and complexity of the implementing circuits. This, of course, depends on the type of application and image standard (e.g. NTSC and PAL). For example, we can observe in Table 4.1 that the multirate SC implementation of a chrominance separation 2-D filter for NTSC systems requires only one fourth of the size of the delay-line (DL) memory block needed in a traditional (non-multirate) design, besides eliminating 42 clock signals and reducing the clock rate from 14 MHz to 3.5 MHz. Similar gains are obtained for the case of the picture enhancement 2-D filter also given in Table 4.1.

Because of their discrete-time nature, the design of SC filters is quite often based on the same methodologies employed for digital filters. In the case of 2-D multirate filters, however, current digital implementations [4.3 – 4.10] based on polyphase filter-banks are not entirely satisfactory because of the complexity to carry-out the 2-D polyphase decomposition and then realizing the analysis and synthesis subband filters. Thus, in order to overcome such difficulties, we investigate in this chapter an alternative technique for the implementation of 2-D decimation filters based on an new form of representing their z-transfer functions. Because a given z-transfer function can be alternatively written in a variable coefficient form, this will be called the *polyphase-coefficient* structure of a 2-D decimation filter.

Table 4.1
Typical requirements of multirate and traditional (non-multirate) 2-D filtering for NTSC systems

Applications of 2D-Filter	Traditional (non-multirate) SC Filtering			Multirate SC Filtering		
	DL Size (cells)	Number of Clocks	Clock (MHz)	DL Size (cells)	Number of Clocks	Clock (MHz)
Chrominance Separation [4.1]	910 x 2	80*	14	230 x 2	38*	3.5
Picture Detail Enhancement [4.2]	910 x 2	80*	14	310 x 2	45*	4

* the number of clocks include those to implement the DL memory block.

For the sake of clarity, we start by describing in Section 4.2 the simpler polyphase-coefficient structure for 1-D decimation filters with both finite and infinite impulse responses. Then, in Section 4.3, we extend such representation also to the case of 2-D decimation filters, also including both FIR and IIR z-transfer functions. The chapter is summarized in Section 4.4.

4.2 MODIFIED 1-D DECIMATION POLYPHASE STRUCTURES

4.2.1 Direct-Form Polyphase-Coefficient FIR Structures

As described in the previous chapter, the prototype filter of a linear, causal and shift-invariant FIR decimator is usually represented by the z-transfer function

$$(4.1) \quad H(z) = \sum_{n=0}^{N-1} h(n) z^{-n},$$

where the unit delay period relates to the input sampling frequency MF_s and N is the length of prototype filter. For simplicity of explanation we consider that $N = ML$, where M is the decimation factor and L an integer greater than 1. In such case, the z-transfer function can also be expressed as

Multirate Switched-Capacitor Circuits for 2-D Signal Processing 61

(4.2) ... $$H(z) = \sum_{m=0}^{M-1} G_m(z) z^{-m},$$

where

(4.3) ... $$G_m(z) = \sum_{i=0}^{L-1} h(m + iM) z^{-iM}.$$

From expressions (4.2) and (4.3) we readily arrive at the direct-form polyphase decimating structure of Fig. 4.1 whose M polyphase sub-filters, all with length L, are represented by the z-transfer functions $F_s(z)$ [4.11 – 4.12]. Because in this structure the filter is first decomposed into M subband filters and then they are added together at the output, this is called the polyphase subband structure of a 1-D FIR decimation filter.

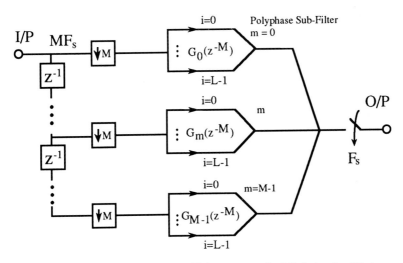

Fig. 4.1. 1-D direct-form polyphase-coefficient structure for 1-D decimation filtering.

4.2.2 ADB Polyphase-Coefficient FIR Structures

An alternative form of implementing the polyphase structure of an FIR decimator is based on the concept of active-delayed blocks (ADB) [4.13]. Here, the prototype filter is decomposed into L blocks, each of which is organized as an M-fold polyphase structure with M single-term coefficients, yielding the z-transfer function

$$(4.4) \quad H(z) = \sum_{i=0}^{L-1}\left[\sum_{m=0}^{M-1} h(m+iM)z^{-m}\right](z^M)^{-i} = \sum_{i=0}^{L-1} \overline{h}_i (z^M)^{-i},$$

where $(z^M)^{-1}$ is the *delay-block* associated with the decimation factor of M and $\overline{h}_i$ are defined as the *polyphase-coefficients* given by

$$(4.5) \quad \overline{h}_i = \sum_{m=0}^{M-1} h(m+iM)z^{-m}.$$

This leads to the ADB polyphase-coefficient structure illustrated in Fig. 4.2. By comparing equations (4.4) and (4.1) we observe that the ADB polyphase-coefficient representation has the same structure as the representation of the prototype z-transfer function and therefore can be implemented using a similar structure. This similarity between the prototype z-transfer function and the resulting decimation z-transfer function will be the basis for deriving the efficient 2-D decimation structures described later.

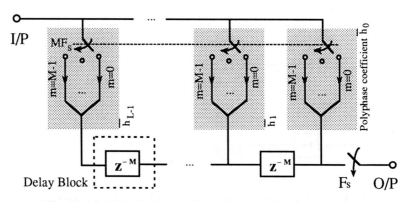

Fig. 4.2. 1-D ADB polyphase-coefficient structure for decimation filtering.

4.2.3 ADB Polyphase-Coefficient IIR Structures

Following a similar procedure as above we start by considering the general z-transfer function of a prototype 1-D IIR filter expressed as

(4.6) $\quad H(z) = \dfrac{N(z)}{D(z)} = \dfrac{\sum_{i=0}^{N-1} a_i z^{-i}}{1 + \sum_{i=1}^{D} b_i z^{-i}} = \dfrac{a_0 + a_1 z^{-1} + \ldots + a_{N-1} z^{-(N-1)}}{1 + b_1 z^{-1} + \ldots + b_{N-1} z^{-(N-1)}}$,

where, for simplicity of explanation, we assume that the orders of the numerator and denominator are equal, i.e. $D = N - 1$. Thus, the above expression can also be written in the form

(4.7) $\quad H(z) = \dfrac{N(z)}{D(z)} = A z^{D-N-1} \dfrac{\prod_{i=1}^{N-1}(z - z_i)}{\prod_{i=1}^{D}(z - p_i)}$,

where z_i and p_i, respectively, are the zeros and poles of $H(z)$ and A is a gain factor. In order to derive the decimating filter function from (4.7) we apply the multirate transformation [4.11 – 4.12]

(4.8) $\quad z^M - p_i^M = (z - p_i)(z^{M-1} + p_i z^{M-2} + \ldots + p_i^{M-1})$,

yielding the polyphase-coefficient form [4.14 – 4.15]

(4.9) ...

$H(z) = \dfrac{\bar{A}_0(z) + \bar{A}_1(z)(z^M)^{-1} + \bar{A}_2(z)(z^M)^{-2} + \ldots + \bar{A}_{N-1}(z)(z^M)^{-(N-1)}}{1 + B_1(z^M)^{-1} + B_2(z^M)^{-2} + \ldots + B_{N-1}(z^M)^{-(N-1)}}$,

where the numerator polyphase-coefficients $\bar{A}_0(z), \bar{A}_1(z), \ldots, \bar{A}_{N-1}(z)$ are given by

(4.10-a) $\quad \bar{A}_0(z) = A_0 + A_1 z^{-1} + \ldots + A_{M-1} z^{-(M-1)}$,

(4.10-b) $\quad \bar{A}_1(z) = A_M + A_{M+1} z^{-1} + \ldots + A_{2M-1} z^{-(M-1)}$,

(4.10-c) $\quad \bar{A}_{N-1}(z) = A_{(N-1)M}$, in the case of $D = N - 1$.

The denominator coefficients $B_1, B_2, \ldots, B_{N-1}$ in (4.9) as well as the coefficients A_0, $A_1, \ldots, A_{(N-1)M}$ in (4.10) are constants. By combining (4.9) and (4.10) we arrive at the more compact expression

$$(4.11) \ldots H(z) = \frac{\sum_{n=0}^{N-1} \left[\sum_{m=0}^{M-1} A_{m+nM} z^{-m} \right] (z^M)^{-n}}{1 + \sum_{n=1}^{N-1} B_n (z^M)^{-n}} = \frac{\sum_{n=0}^{N-1} \bar{A}_n (z^M)^{-n}}{1 + \sum_{n=1}^{N-1} B_n (z^M)^{-n}}$$

to represent the z-transfer function of an IIR decimating filter with decimating factor M. Given the similarity between expressions (4.11) and (4.6), the IIR decimating filter can be realized by combining the type of ADB polyphase-coefficient structure of Fig. 4.2 together with a classical transposed direct-form recursive network, and thus yielding the schematic representation of Fig. 4.3 [4.14 – 4.15]. The delay-blocks $(z^M)^{-1}$ are the common elements in both structures.

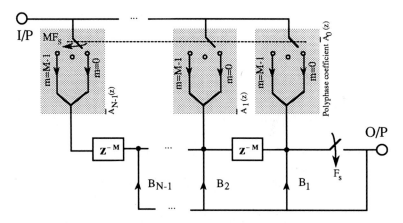

Fig. 4.3. 1-D ADB polyphase-coefficient structure for IIR decimation filters.

4.3 POLYPHASE-COEFFICIENT STRUCTURES FOR 2-D DECIMATION FILTERS

4.3.1 FIR Filtering

The general z-transfer function of a 2-D linear, shift-invariant FIR filter can be expressed as

Multirate Switched-Capacitor Circuits for 2-D Signal Processing

$$(4.12) \quad H(z_1, z_2) = \sum_{n_2=0}^{N_2-1} \sum_{n_1=0}^{N_1-1} h(n_1, n_2) z_1^{-n_1} z_2^{-n_2},$$

where N_1 and N_2, respectively, represent the prototype filter length in the z_1 and the z_2 dimensions. The z_1^{-1} DL term refers to the input sampling frequency of $M_1 F_{s1}$ along the vertical axis whereas the z_2^{-1} unit-delay term refers to the input sampling frequency of $M_2 F_{s2}$ along the horizontal axis. There are two possibilities for the implementation of such decimation filtering functions. In one case, decimation is implemented in only one dimension, while in the other case decimation is implemented in both dimensions. In general, when decimation occurs in both dimensions the above prototype expression (4.12) can be rewritten as

$$(4.13) \quad H(z_1, z_2) = \sum_{n_1=0}^{L_1-1} \left[\sum_{m_1=0}^{M_1-1} h_1(m_1 + n_1 M_1) z_1^{-m_1} \right] \left(z_1^{M_1} \right)^{-n_1},$$

where

$$(4.14)$$

$$h_1(m_1 + n_1 M_1) = \sum_{n_2=0}^{L_2-1} \left[\sum_{m_2=0}^{M_2-1} h(m_1 + n_1 M_1, m_2 + n_2 M_2) z_2^{-m_2} \right] \left(z_2^{M_2} \right)^{-n_2}.$$

In the above equations, M_1 and M_2 are the decimation factors whereas L_1 and L_2 represent, respectively, the number of sub-filters in both dimensions. Similarly to the 1-D case discussed before, we can define the 2-D polyphase-coefficient

$$(4.15) \quad \bar{h}_{i,j} = \sum_{m_1=0}^{M_1-1} \sum_{m_2=0}^{M_2-1} h(m_1 + n_1 M_1, m_2 + n_2 M_2) z_2^{-m_2} z_1^{-m_1}$$

as a function of both decimating factors. This, in turn, yields the more compact expression with 2-D polyphase-coefficient form

$$(4.16) \quad H(z_1, z_2) = \sum_{n_1=0}^{L_1-1} \sum_{n_2=0}^{L_2-1} \bar{h}_{i,j} \left(z_2^{M_2} \right)^{-n_2} \left(z_1^{M_1} \right)^{-n_1},$$

where the subscripts (m_1, n_1) denote the m_1th impulse response coefficient in the n_1th sub-filter in the vertical dimension, and the subscripts (m_2, n_2) represent the

m_2th impulse response coefficient in the n_2th sub-filter in the horizontal dimension. Therefore, one 2-D impulse response coefficient in such structure must be expressed by four parameters: for example, the coefficient $h(m_1 + n_1M_1, m_2 + n_2M_2)$ corresponds to a coefficient between the (m_1, n_1)th sub-filter in z_1 and the (m_2, n_2)th sub-filter in z_2.

The above z-transfer function (4.16) can be implemented using the 2-D direct-form decimation polyphase-coefficient structure of Fig. 4.4, comprising $(z_1^{M_1})^{-1}$ and $(z_2^{M_2})^{-1}$ delayed-blocks. The horizontal $(z_2^{M_2})^{-1}$ delayed-block is realized as in the type of ADB structures described before. The vertical delayed-block, $(z_1^{M_1})^{-1}$, is associated with the DL memory blocks and its implementation will be described later.

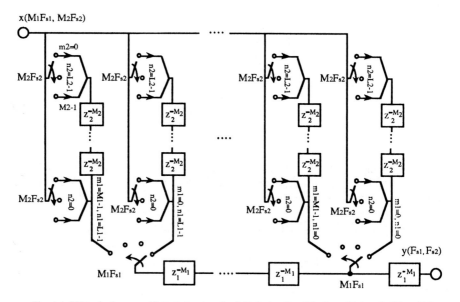

Fig. 4.4. FIR polyphase-coefficient structure for 2-D decimation filtering with length ($N_1 \times N_2$).

4.3.2 IIR Filtering

The general z-transfer function of a linear, shift-invariant IIR 2-D filter can be expressed as the expression (4.17) where, as before, z_2^{-1} and z_1^{-1} are the unit-delay periods referring to the input sampling frequencies M_2F_{s2} and M_1F_{s1}, respectively. When the multirate transformation [4.11 – 4.12] is applied in both dimensions of the above z-transfer function the order of the numerator of the modified decimating filter function increases proportionally to $(M_1M_2)^2$, and thus becomes too complex for practical implementation.

$$(4.17) \ldots H(z_1, z_2) = \frac{N(z_1, z_2)}{D(z_1, z_2)} = \frac{\sum_{n_1=0}^{N_1-1}\sum_{n_2=0}^{N_2-1} a_{n_1,n_2} z_2^{-n_2} z_1^{-n_1}}{1 + \sum_{\substack{n_1=0 \\ n_1+n_2 \neq 0}}^{N_1-1}\sum_{n_2=0}^{N_2-1} b_{n_1,n_2} z_2^{-n_2} z_1^{-n_1}},$$

For the sake of practicality, we therefore restrict this study to separable denominator polynomials, i.e. such that $D(z_1, z_2) = D_1(z_1)D_2(z_2)$. In this case, the above expression can be rewritten as

$$(4.18) \ldots H(z_1, z_2) = \frac{\sum_{n_1=0}^{N_1-1}\sum_{n_2=0}^{N_2-1} a_{n_1,n_2} z_2^{-n_2} z_1^{-n_1}}{(1 + \sum_{n_1=1}^{N_1-1} b_{n_1} z_1^{-n_1})(1 + \sum_{n_2=1}^{N_2-1} b_{n_2} z_2^{-n_2})}.$$

For multirate implementation two cases can be considered. In one case, the multirate transformation is applied to only one of either dimensions whereas in the other case the multirate transformation is applied in both dimensions. Considering the latter more general case, and following the procedure previously described, the z-transfer function for the 2-D IIR decimation filter can be expressed as

(4.19) ...

$$H(z_1, z_2) = \frac{(\sum_{n_1=0}^{N_1-1}\sum_{n_2=0}^{N_2-1} a_{n_1,n_2} z_2^{-n_2} z_1^{-n_1})(\sum_{k_1=0}^{(M_1-1)(N_1-1)} p_{k_1} z_1^{-k_1})(\sum_{k_2=0}^{(M_2-1)(N_2-1)} q_{k_2} z_2^{-k_2})}{(1 + \sum_{n_1=1}^{N_1-1} b_{n_1} z_1^{-n_1})(1 + \sum_{n_2=1}^{N_2-1} b_{n_2} z_2^{-n_2})(\sum_{k_1=0}^{(M_1-1)(N_1-1)} p_{k_1} z_1^{-k_1})(\sum_{k_2=0}^{(M_2-1)(N_2-1)} q_{k_2} z_2^{-k_2})}$$

where the coefficients p_{k1} and q_{k2} are determined in a similar way as explained before for the 1-D IIR case. After defining the 2-D polyphase-coefficient as

$$(4.20) \quad \bar{A}_{n_1, n_2} = \sum_{m_1=0}^{M_1-1} \sum_{m_2=0}^{M_2-1} A(m_1 + n_1 M_1, m_2 + n_2 M_2) z_2^{-m_2} z_1^{-m_1},$$

and carrying out simple algebraic manipulations we arrive at the more compact expression

$$(4.21) \quad H(z_1, z_2) = \frac{\sum_{n_1=0}^{N_1-1} \sum_{n_2=0}^{N_2-1} \bar{A}_{n_1, n_2} \left(z_2^{M_2}\right)^{-n_2} \left(z_1^{M_1}\right)^{-n_1}}{\left[1 + \sum_{n_1=1}^{N_1-1} B_{n_1}(z_1^{M_1})^{-n_1}\right]\left[1 + \sum_{n_2=1}^{N_2-1} B_{n_2}(z_2^{M_2})^{-n_2}\right]}$$

to describe the z-transfer function of the 2-D IIR decimation filter in polyphase-coefficient form.

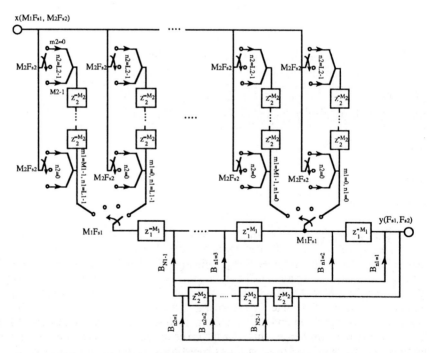

Fig. 4.5. Polyphase-coefficient structure for 2-D IIR decimation filters with length ($N_1 \times N_2$).

From the above expression, we can readily obtain the corresponding decimation architecture shown in Fig. 4.5 where it can be seen the combination of a 2-D ADB polyphase structure realizing the numerator polynomial function in (4.21), as in the 2-D FIR case, together with 2-D recursive network realizing the denominator polynomial function in (4.21).

A simplified version of the above structure can be obtained assuming that, besides the denominator, the numerator function is also separable. This corresponds to an original prototype filter function given by

$$(4.22) \quad H(z_1, z_2) = H_1(z_1) H_2(z_2) = \frac{N_1(z_1) N_2(z_2)}{D_1(z_1) D_2(z_2)},$$

and which, in turn, leads to the modified decimating filter function expressed as

$$(4.23) \quad H(z_1, z_2) = \frac{\left[\sum_{n_1=0}^{N_1-1} \bar{A}_{n_1}(z_1^{M_1})^{-n_1} \right] \left[\sum_{n_2=0}^{N_2-1} \bar{A}_{n_2}(z_2^{M_2})^{-n_2} \right]}{\left[1 + \sum_{n_1=1}^{N_1-1} B_{n_1}(z_1^{M_1})^{-n_1} \right] \left[1 + \sum_{n_2=1}^{N_2-1} B_{n_2}(z_2^{M_2})^{-n_2} \right]}.$$

The corresponding 2-D decimating filter can now be implemented as a cascaded of two 1-D decimating filters, each of which, in turn, can be implemented using the basic polyphase-coefficient structures derived in the previous sections.

4.4 SUMMARY

The implementation of 2-D SC filtering functions using multirate processing techniques is motivated by the need to reduce both the processing speed and complexity of the implementing circuits, notably the DL memory blocks. However, unlike in many other applications of SC filters, current digital implementations based on polyphase filter-banks are not entirely satisfactory for emulation by their SC counterparts because of the complexity to carry out the 2-D polyphase decomposition and then realizing analysis and synthesis subband filters. To overcome this limitation, this chapter investigated alternative solutions for the implementation of 2-D decimation filters based on a modified representation of their z-transfer functions using *polyphase-coefficients*. For finite impulse response z-transfer functions the resulting polyphase-coefficient structures can be implemented either in direct-form or using active-delayed blocks. Infinite impulse response z-transfer functions are best implemented using a combination of ADB polyphase-coefficient structures, for the numerator, together with a direct-form recursive structure for the denominator. This, however, is limited to the practical case where the denominator polynomial function is separable, i.e. can be expressed by the product of two independent functions, one for each filter dimensions. The SC implementation of such structures, for both FIR and IIR 2-D decimation filters will be addressed in the next chapter.

REFERENCES

[4.1] K. A. Nishimura and P. R. Gray, "A Monolithic Analog Video Comb Filter in 1.2-µm CMOS", *IEEE Journal of Solid-State Circuits*, Vol. 28, No. 12, pp. 1331-1339, December 1993.

[4.2] K.Matsui, T.Matsuura, S.Fukasawa, Y.Izawa, Y.Toba, N.Miyake and K.Nagasawa, "CMOS Video Filters Using SC 14-MHz Circuits", *IEEE Journal of Solid-State Circuits*, Vol. sc-20, No. 6, pp. 1096-1101, December 1985.

[4.3] P.P. Vaidyanathan, *"Multirate Systems and Filter Banks"*, Prentice-Hall, Englewood Cliffs, NJ, 1993.

[4.4] T. C. Chen and R. J. P. De Figueiredo, "Image Decimation and Interpolation Based on Frequency Domain Analysis", *IEEE Transactions on Communications*, Vol. com-32, No. 4, pp. 479-484, April 1984.

[4.5] M. Renfors, "Multi-Dimensional Sampling Structure Conversion with 1-D Nth-Band Filters", *IEEE International Symposium on Circuits and Systems*, pp. 1502-1506, June 1989.

[4.6] Q.S. Gu, M.N.S. Swamy, L.C.K. Lee and M.O. Ahmad, "IIR Digital Filters for Sampling Structure Conversion and Deinterlacing of Video Signals", *IEEE International Symposium on Circuits and Systems*, pp. 973-976, San Diago, May 1995.

[4.7] G. Schamel, "Pre- and Post-filtering of HDTV Signals for Sampling Rate Reduction and Display Up-conversion", *IEEE Transactions on Circuits and Systems*, vol. 34, no.11, pp. 1432-1439, November 1987.

[4.8] P. Siohan, "2-D FIR Filter Design for Sampling Structure Conversion", *IEEE Transactions on Circuits and Systems for Video Technology*, vol. 1, no.4, pp. 337-350, December 1991.

[4.9] R. Ansari and C.L. Lau, "2-D IIR Filters for Exact Reconstruction in Tree-Structured Sub-Band Decomposition", *Electronics Letters*, vol. 23, No.12, pp. 633-634, 4th June 1987.

[4.10] U. Pestel and K. Gruger, "Design of HDTV Subband Filterbanks Considering VLSI Implementation Constraints", *IEEE Transactions on Circuits and Systems for Video Technology*, vol. 1, no.1, pp. 14-21, March 1991.

[4.11] Maurice Bellanger, *"Digital Processing of Signals"*, Academic Press, 1978.

[4.12] R. Crochiere and L.R. Rabiner, *"Multirate Digital Signal Processing"*, Prentice-Hall, Englewood Cliffs, NJ, 1983.

[4.13] J. E. Franca and S. Santos, "FIR Switched-Capacitor Decimators with Active-Delayed Block Polyphase Structures", *IEEE Transactions on Circuits Systems*, vol. 35, no. 8, pp. 1033-1037, August 1988.

[4.14] R.P. Martins and J.E. Franca, "A 2.4-µm CMOS SC Video Decimator with Sampling Rate Reduction from 40.5 MHz to 13.5 MHz", *IEEE Custom Integrated Circuits Conference*, pp. 1-4, San Diego, May 1989.

[4.15] W. Ping and J. E. Franca, "New Form of Realization of IIR Switched-Capacitor Decimators", *Electronics Letters*, vol. 29, No.11, pp. 953-954, 27th May 1993.

5

SC ARCHITECTURES FOR 2-D DECIMATION FILTERS IN POLYPHASE-COEFFICIENT FORM

5.1 INTRODUCTION

In the previous chapter we introduced the polyphase-coefficient structures for the efficient implementation of 2-D decimation filters. In this chapter, we shall discuss the realization of such structures using SC circuit techniques. First, in Section 5.2, we examine the basic SC building blocks needed for multirate implementation. Then, in Section 5.3, we describe SC architectures that are particularly suitable for the implementation of 2-D decimation filters described in polyphase-coefficient form, both with finite and infinite impulse responses and with different filter lengths and decimation ratios in both dimensions. The design of such SC architectures is then illustrated considering the examples of an FIR 2-D filter, given in Section 5.4, and also of an IIR 2-D filter, described in Section 5.5. Finally, the chapter is summarized in Section 5.6.

5.2 SC BUILDING BLOCKS

In this section and thereafter, the analog switches needed for the realization of SC circuits will be represented by a small square with an inserted number

74   5. SC Architectures for 2-D Decimation Filters in Polyphase-Coefficient Form

indicating the clock phase of a given switch timing that controls its operation. This is illustrated in Fig. 5.1 where we can observe various types of basic building blocks, both quasi-passive, i.e. containing only switches and capacitors, and active, and the corresponding relationships between input and output variables [5.1 – 5.9].

Referring to the switch timing diagram indicated in Fig. 5.1(g), the quasi-passive SC networks of Fig. 5.1(a), (b), (c) and (d) operate as delayed-coefficient-multipliers with different delays and coefficients determined by the controlling clock phases. The active SC circuit shown in Fig. 5.1(e) realizes a Backward-Euler integrator, whereas the circuit of Fig. 5.1(f) gives the example of a 3-fold decimator. Here, the output sample produced in phase 1 is determined by the three input samples taken by the input SC branches during phases 1, 2, and 3.

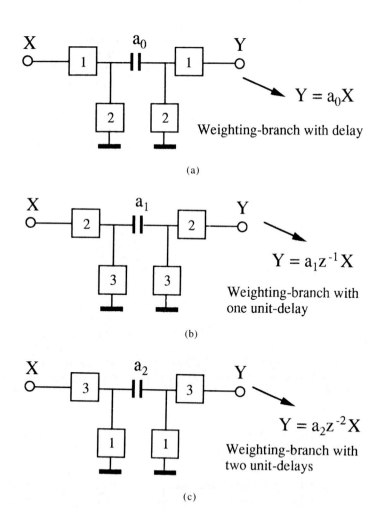

$Y = a_0 X$

Weighting-branch with delay

(a)

$Y = a_1 z^{-1} X$

Weighting-branch with one unit-delay

(b)

$Y = a_2 z^{-2} X$

Weighting-branch with two unit-delays

(c)

Multirate Switched-Capacitor Circuits for 2-D Signal Processing

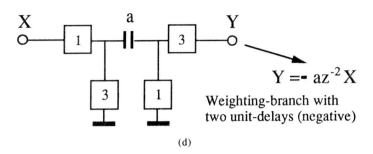

(d) Weighting-branch with two unit-delays (negative)
$$Y = -az^{-2}X$$

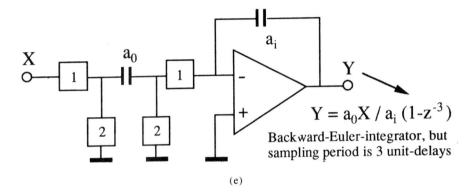

(e) Backward-Euler-integrator, but sampling period is 3 unit-delays
$$Y = a_0 X / a_i (1-z^{-3})$$

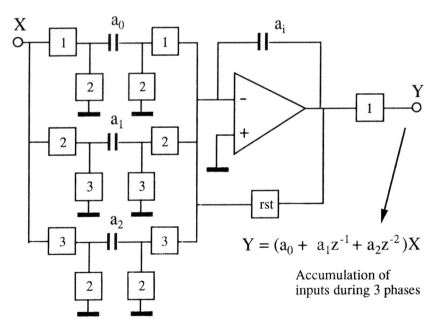

(f) Accumulation of inputs during 3 phases
$$Y = (a_0 + a_1 z^{-1} + a_2 z^{-2})X$$

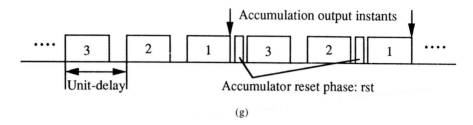

(g)

Fig. 5.1. Basic SC building blocks and switch timing. (a) Weighting branch without delay. (b) Weighting branch with one unit-delay. (c) Weighting branch with two unit-delays. (d) Weighting branch with two negative unit-delays. (e) Backward-Euler integrator with three unit-delay sampling periods. (f) Accumulation block to accumulate input samples during three phases. (g) Switch timing.

5.3 2-D POLYPHASE-COEFFICIENT SC STRUCTURES

As can be seen from expressions (4.13) and (4.21) derived in Chapter 4, the complexity of a 2-D decimation filter basically depends on the filtering orders N_1 and N_2 and decimation factors M_1 and M_2, respectively in the z_1 and z_2 dimensions. To explain the derivation of the corresponding SC structures we consider the example of FIR 2-D decimation filters with four different levels of complexity: (i) $N_1 + 1 \leq M_1$ and $N_2 + 1 \leq M_2$, (ii) $N_1 + 1 \leq M_1$ and $N_2 + 1 > M_2$, (iii) $N_1 + 1 > M_1$ and $N_2 + 1 \leq M_2$, and (iv) $N_1 + 1 > M_1$ and $N_2 + 1 > M_2$. In all four cases we consider, for simplicity, a small size 2-D input signal in which the scanning lines contain 24 pixels each and are decimated 3-fold both in the z_1 and z_2 dimensions, i.e. $M_1 = M_2 = 3$.

5.3.1 Case 1: $N_1 + 1 \leq M_1$ and $N_2 + 1 \leq M_2$

Besides $M_1 = M_2 = 3$ indicated above, this case is explained considering a 2nd-order filtering in both dimensions, i.e. $N_1 = N_2 = 2$. From expression (4.13) and Fig. 4.4, we obtain the 2-D structure shown in Fig. 5.2(a). Then, by using the SC building blocks introduced before, we derive the SC structure of Fig. 5.2(b). The associated switch timing of Fig. 5.2(c) comprises three sets of waveforms: switching phases 1, 2 and 3, sample the input pixels while switching phases a1, a2, a3, a4, a5, a6, a7 and a8 determine the decimated output pixels and switching phases L0, L1 and L2 define the line sampling. In all three polyphase structures the input switches controlled by switching phases 1, 2, 3 sequentially sample the 2-D input signal $V_i(z_1, z_2)$ into capacitors C_{lp}, where $l = 0, 1, 2$ is the line sampling index and $p = 0, 1, 2$ is the input pixel sampling index.

Multirate Switched-Capacitor Circuits for 2-D Signal Processing 77

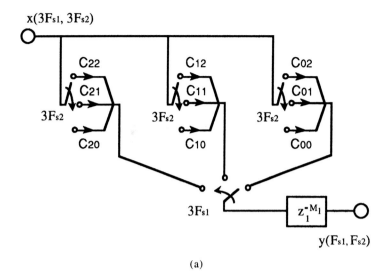

(a)

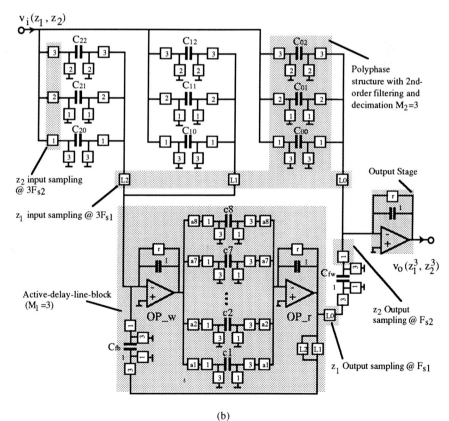

(b)

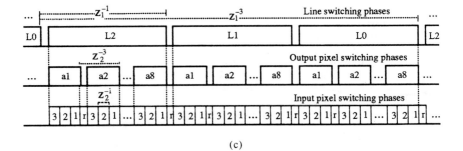

(c)

Fig. 5.2. An illustrative example of a 2-D SC filter circuit with 2nd-order filtering, $N_1 = N_2 = 2$, and decimation of $M_1 = M_2 = 3$. (a) Architecture, (b) schematic diagram, and (c) associated switch timing.

The output packets of charge representing the decimated pixels from each polyphase structure, i.e. produced during phase 1 at the end of every switching phases a1, a2, a3, a4, a5, a6, a7, and a8, can be expressed as

$$(5.1) \quad \Delta Q_{lo}(z_1, z_2^3) = \sum_{p=0}^{2} V_i(z_1, z_2) C_{lp} z_2^{-p}.$$

In the above expression (5.1) $o = 1, 2, 3, ... 8$ represents the decimated pixel index and z_2^{-1} is the delay with input sampling period at $3F_{s2}$. In the z_1 dimension, the switching phases L2, L1 and L0 periodically switch lines into the active DL block where they are accumulated and decimated, as shown respectively in Fig. 5.3(a), Fig. 5.3(b) and Fig. 5.3(c).

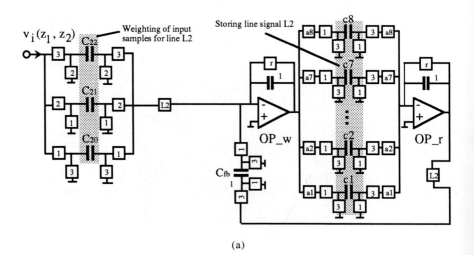

(a)

Multirate Switched-Capacitor Circuits for 2-D Signal Processing 79

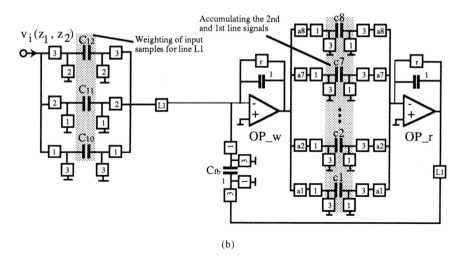

(b)

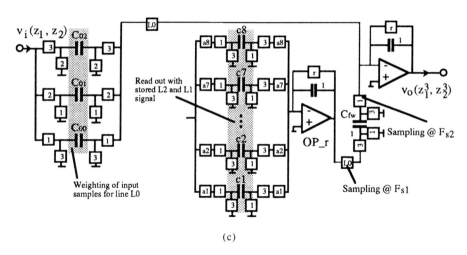

(c)

Fig. 5.3. Unfolded operation of the SC circuit of Fig. 5.2 during (a) switching phase L2, (b) switching phase L1 and (c) switching phase L0.

For the switching phase L2, shown in Fig. 5.3(a), the weighted charge packets from the associated polyphase structure enter the accumulating block OP_w where they are added sequentially with the charge packets from the feedback branch Cfb yielding, during each phase 1, the output charge packets ΔQ_{21}, ΔQ_{22}, ΔQ_{23}, ... ΔQ_{28} that are memorized into the DL block formed by capacitors C1 to C8. Similar operations are performed during the switching phase L1 shown in Fig. 5.3(b). In this case, however, it should be noticed that before the output charge packets ΔQ_{11}, ΔQ_{12}, ΔQ_{13}, ... ΔQ_{18} enter the DL during phase 1, during each phase 3 the previously stored charge packets ΔQ_{21}, ΔQ_{22}, ΔQ_{23}, ... ΔQ_{28} have been sequentially read out by OP_r and feedback to OP_w to accumulate with the charge packets from line 1. The output charge packets of OP_w, which are now $(\Delta Q_{21}+\Delta Q_{11})$, $(\Delta Q_{22}+\Delta Q_{12})$, ... $(\Delta Q_{28}+\Delta Q_{18})$, are then memorized into capacitors C1 to C8. Finally, when the switching phase L0 becomes active, Fig. 5.3(c), the weighted charge packets from the associated polyphase structure directly go to the output stage to produce the weighted and decimated charge packets ΔQ_{01}, ΔQ_{02}, ΔQ_{03}, ... ΔQ_{08}. Meanwhile, in the active DL block, since the feedback branch is replaced by the feedforward branch Cfw with the switching line phase L0, the stored charge packets representing the two previous lines are read sequentially out to the output stage instead of feeding back to the OP_w input. Hence, at the end of each phase 1 of L0, the charge packets at the output stage can be expressed as

$$(5.2) \quad \Delta Q_o(z_1^3, z_2^3) = \sum_{l=0}^{2} \Delta Q_{lo}\left(z_1, z_2^3\right) z_1^{-l}$$

After substituting (5.1) into (5.2) it results

$$(5.3) \quad \Delta Q_o(z_1^3, z_2^3) = \sum_{l=0}^{2} \sum_{p=0}^{2} V_i\left(z_1, z_2\right) C_{lp} z_2^{-p} z_1^{-l}$$

representing the relationship between the input signal sampled in phases 1, 2 and 3 and the decimated output charge packets at the end of phase L0. From these, we finally obtain at the output of the decimation filter the sampled output voltage described by

$$(5.4) \quad V_o(z_1^3, z_2^3) = \sum_{l=0}^{2} \sum_{p=0}^{2} V_i(z_1, z_2) C_{lp} z_2^{-p} z_1^{-l}$$

where every memory capacitor is assumed equal to one.

5.3.2 Case 2: $N_1 + 1 \leq M_1$ and $N_2 + 1 > M_2$

When the filter order increases in the z_2 dimension, such that $N_2 + 1 > M_2$, the polyphase structures are more efficiently implemented using the type of 1-D ADB architectures described in [5.10]. This is illustrated in Fig. 5.4 considering the SC implementation of a 2-D FIR decimator with $N_2 = 3$, $N_1 = 2$ and $M_2 = M_1 = 3$.

Since only the length in the z_2 dimension has increased we can easily derive the equations describing the operation of the circuit following the procedures discussed above. Hence, at the outputs of each one of the ADBs the charge packets of the decimated pixel are given by

$$(5.5) \quad \Delta Q_{lo}(z_1, z_2^3) = \sum_{p=0}^{2} V_i(z_1, z_2) C_{lp} z_2^{-p} + V_i(z_1, z_2) C_{l3} z_2^{-3}$$

Furthermore, since the structure in the z_1 dimension remains the same as in Fig. 5.2(a), we can replace (5.5) into (5.2) yielding the expression

$$(5.6) \quad \Delta Q_o(z_1^3, z_2^3) = \sum_{l=0}^{2} \left(\sum_{p=0}^{2} V_i(z_1, z_2) C_{lp} z_2^{-p} + V_i(z_1, z_2) C_{l3} z_2^{-3} \right) z_1^{-l}$$

that represents the output decimated and filtered charge packets in both dimensions. By sampling the output terminal of the output stage, during the line-decimation output phase L0, we obtain

$$(5.7) \quad V_o(z_1^3, z_2^3) = \sum_{l=0}^{2} \left(\sum_{p=0}^{2} V_i(z_1, z_2) C_{lp} z_2^{-p} + V_i(z_1, z_2) C_{l3} z_2^{-3} \right) z_1^{-l}$$

describing the SC 2-D decimation filtering of Fig. 5.4.

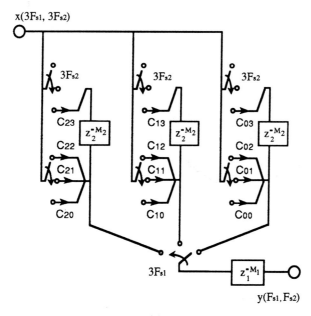

(a)

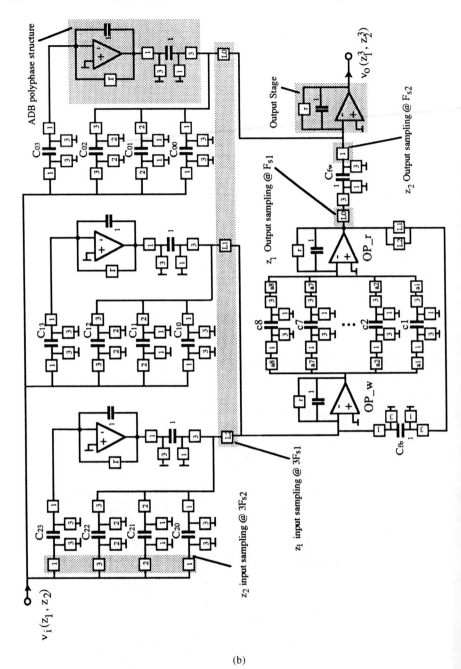

(b)

Fig. 5.4. Example of a 2-D SC FIR decimation filter with $N_1=2$, $N_2=3$ and $M_1=M_2=3$. (a) Architecture and (b) schematic diagram.

5.3.3 Case 3: $N_1 + 1 > M_1$ and $N_2 + 1 \leq M_2$

When the filtering order increases in the z_1 dimension it is necessary to introduce additional active DL blocks, as illustrated in Fig. 5.5 for the example of a 2-D FIR decimator with $N_1 = 4$, $N_2 = 2$ and $M_2 = M_1 = 3$. The basic SC architecture is given in Fig. 5.5(a) and the corresponding SC circuit and associated switch timing are both shown in Fig. 5.5(b). Here, besides the operations previously described there is an additional operation concerning the charge transfer from the 2nd active DL block to the 1st active DL block during each phase 1 of L0. This will be explained in detail based on the unfolded representation of the various phases of operation shown in Fig. 5.6.

The configurations of Fig. 5.6(a), Fig. 5.6(b) and Fig. 5.6(c), respectively, illustrate the operation during the three line sampling phases L2, L1, L0. During line phase L2, shown in Fig. 5.6(a), we assume that the 1st DL block has already stored the charge packets transferred from the 2nd active DL block in the last line-decimation period and therefore the current charge packets from the associated polyphase structure enter the accumulating block OP_w of the 1st DL block where they are added to the stored charge packets of the two previous lines. When line phase L1 is active, as shown in Fig. 5.6(b), both active DL blocks independently receive the charge packets from the associated polyphase structures yielding at the end of line phase L1 the stored charge packets indicated in each memory capacitor.

During the line-decimation output phase L0, let us first observe the operation shown in Fig. 5.6(c) for the 1st active DL block. At each phase 3 of L0 the stored charge packets are sequentially read out by OP_r with Cfw, and go directly to the output stage where they are added to the charge packets of the associated polyphase structure to produce the final output of this filter.

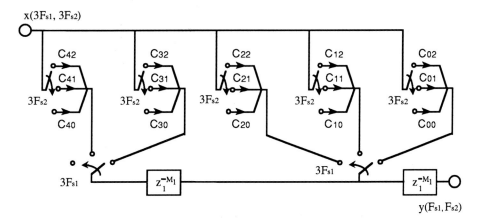

(a)

84   5. SC Architectures for 2-D Decimation Filters in Polyphase-Coefficient Form

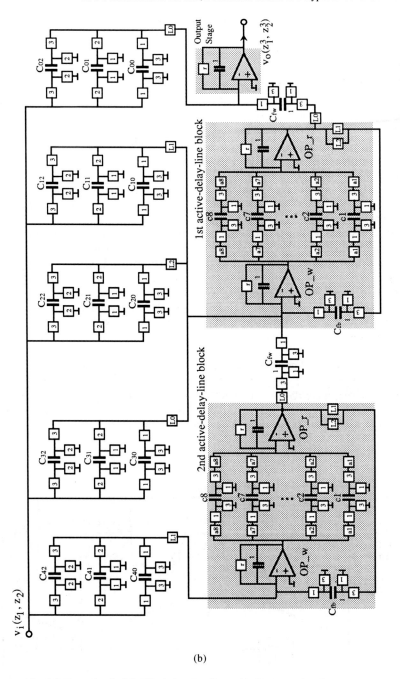

(b)

Fig. 5.5. Example of a 2-D FIR decimation filter with $N_1=4$, $N_2=2$ and $M_1=M_2=3$, (a) Architecture (b) SC implementation.

Just after each phase 3, as we can see from the switch timing diagram of Fig. 5.2(c), the memory capacitors are empty. Then, at each phase 1, the charge packets from the polyphase structure corresponding to the set of capacitors C_{40}, C_{41} and C_{42} are added in the accumulating block OP_w of the 1st delay-line block together with the read-out charge packets from the 2nd active DL block and memorized into the corresponding capacitors. At the end of line phase L0, the memory capacitors of the 2nd active DL block become empty while those of the 1st active DL block take the same initial condition as in Fig. 5.6(a) to start a new line-decimation.

Thus, the sampled voltage at the output of the output stage can be expressed as

(5.8) ...

$$V_o(z_1^3, z_2^3) = \sum_{l=0}^{2} \sum_{p=0}^{2} V_i(z_1, z_2) C_{lp} z_2^{-p} z_1^{-l} + \sum_{l=3}^{4} \sum_{p=0}^{2} V_i(z_1, z_2) C_{lp} z_2^{-p} z_1^{-l}$$

describing the SC 2-D decimation filter of Fig. 5.5.

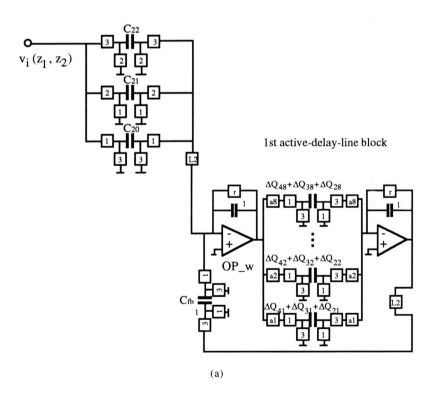

(a)

86 5. SC Architectures for 2-D Decimation Filters in Polyphase-Coefficient Form

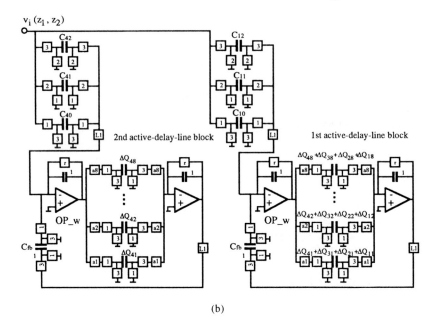

(b)

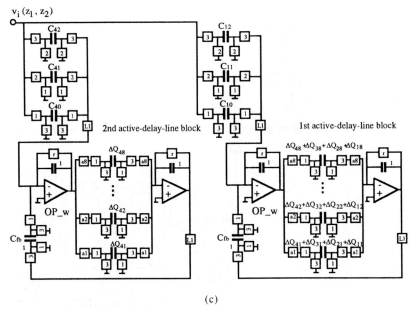

(c)

Fig. 5.6. Unfolded operation of the SC circuit of Fig. 5.5. (a) Switching phase L2. (b) Switching phase L1. (c) Switching phase L0, in which the synchronized operation of charge transfer from the 2nd to the 1st active DL blocks takes place.

5.3.4 Case 4: $N_1 + 1 > M_1$ and $N_2 + 1 > M_2$

For completeness, we illustrate in Fig. 5.7 the SC implementation of a 2-D FIR decimation filter with $N_1 = 4$, $N_2 = 3$ and $M_1 = M_2 = 3$. As before, the basic polyphase-coefficient representation is given in Fig. 5.7(a) and the corresponding SC implementation and associated switch timing are both shown in Fig. 5.7(b). Because of the increased filtering order in both dimensions the circuit includes both 1-D ADB and active DL block polyphase structures.

By combining expressions (5.7) and (5.8) we easily arrive at

(5.9) ...

$$V_o(z_1^3, z_2^3) = \sum_{l=0}^{2} \left(\sum_{p=0}^{2} V_i C_{lp} z_2^{-p} + V_i C_{l3} z_2^{-3} \right) z_1^{-l} +$$

$$+ \left(\sum_{p=0}^{2} V_i C_{3p} z_2^{-p} + V_i C_{33} z_2^{-3} \right) z_1^{-3}$$

representing the SC 2-D decimation filter of Fig. 5.7.

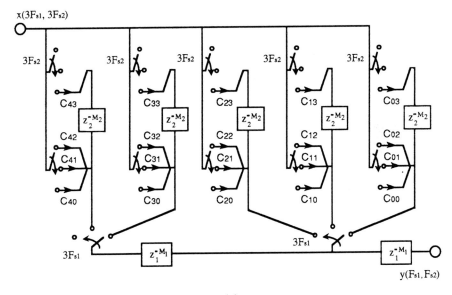

(a)

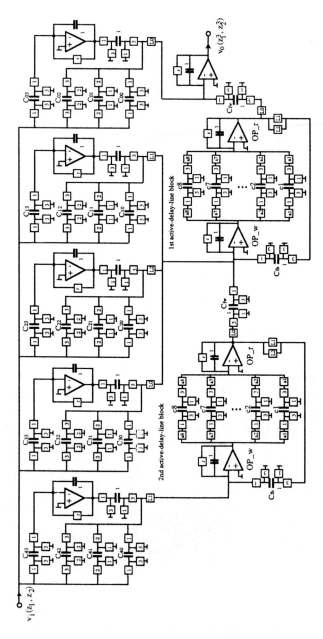

(b)

Fig. 5.7. An example of a 2-D SC FIR decimation filter with $N_1=4$, $N_2=3$ and $M_1=M_2=3$. (a) Architecture. (b) SC implementation.

5.4 DESIGN EXAMPLE OF AN FIR 2-D DECIMATION FILTER

Consider here the implementation of a 2-D FIR filter to meet the specifications indicated in Table 5.1. Because of practical limitations of the simulator SWITCAP2 [5.11] we employ small length DL blocks which prevent simulating the whole frequency scale in the horizontal z_2 dimension according to the typical image frequency specifications. However, the effect of using such a simplified image representation is just proportional to reducing the z_2 frequency scale and hence does not affect the envisaged verification of the 2-D operation of the circuits.

Table 5.1
Specifications of a 2-D FIR filter

Format of the 2-D input signal	24 pixels per scanning-line
Vertical input sampling frequency ($3F_{s1}$)	15 kHz
Vertical decimation factor (M_1)	3
Horizontal input sampling rate ($3F_{s2}$)	24 x 15 kHz = 360 kHz
Horizontal decimation factor (M_2)	3
Prototype filter order	(2 x 2)nd order

From the descriptions in the previous section, and based on the 2-D polyphase-coefficient architecture shown in Fig. 5.1, we obtain the 2-D SC decimation filter schematically illustrated in Fig. 5.8(a), together with the corresponding multiphase switching waveforms indicated in Fig. 5.8(b). The nominal capacitance values of all the input weighting-capacitors are chosen to be equal to the values of their impulse response coefficients, as indicated in Table 5.2. Each group of capacitors, for example, C_{a00}, C_{a01}, C_{a02} and C_{a10}, C_{a11}, C_{a12} are connected to the corresponding line-switches. Only 8 memory capacitors in the active DL block are employed to accumulate and store the previous two line pixels instead of the 24 capacitors that would be needed in the DL block if traditional non-multirate techniques were employed.

Table 5.2
Impulse response coefficients of the 2-D FIR prototype filter function

C_{00} = 4.8634E-4	C_{01} = 7.3510E-3	C_{02} = 4.8634E-4
C_{10} = 7.3510E-3	C_{11} = 1.1111E-1	C_{12} = 7.3510E-3
C_{20} = 4.8634E-4	C_{21} = 7.3510E-3	C_{22} = 4.8634E-4

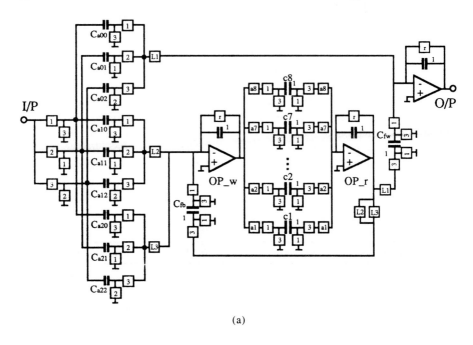

(a)

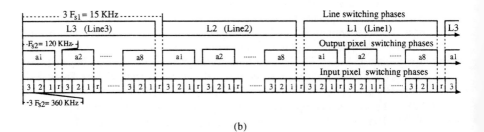

(b)

Fig. 5.8: (a) SC network and (b) controlling switching waveforms for the realization of 2-D FIR decimation filtering function.

The 2-D amplitude response resulting from the simulated operation of the above circuit, using SWITCAP2, is shown in Fig. 5.9(a). Fig. 5.9(b) illustrates the additional sample and hold effect in both z_1 and z_2 dimensions.

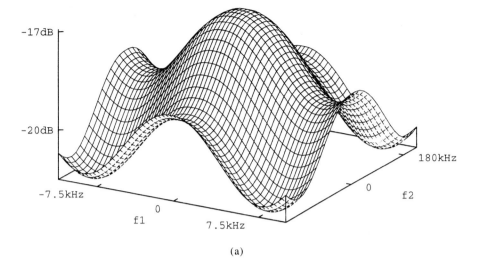

(a)

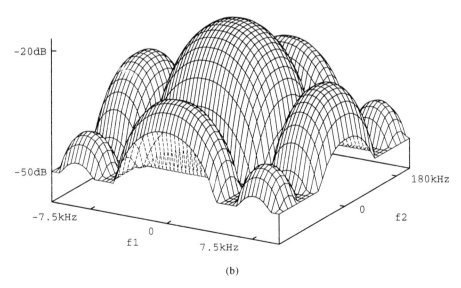

(b)

Fig. 5.9. Computer simulated amplitude responses of the 2-D SC decimating circuit of Fig. 5.8. (a) Impulse sampled. (b) Sampled-and-hold effect in both dimensions.

5.5 DESIGN EXAMPLE OF AN IIR 2-D DECIMATION FILTER

For illustrating the practical SC implementation of a 2-D IIR polyphase-coefficient architecture we consider next the design of a 2-D decimation filter separable in both the numerator and denominator polynomial functions. The target spatial and temporal parameters are given in Table 5.3, and we assume that the original prototype filter coefficients, equal in both dimensions, are those given in Table 5.4.

Table 5.3
Specifications of a 2-D IIR decimating filter

Format of the input 2-D signal	24 pixels per scanning-line
Vertical input sampling rate ($2F_{s1}$)	20 kHz
Vertical decimation factor (M_1)	2
Horizontal input sampling rate ($3F_{s2}$)	24 x 20 kHz = 480 kHz
Horizontal decimation factor (M_2)	3

Table 5.4
Coefficients of the prototype 2-D IIR filter function

Numerator Coefficients	Denominator Coefficients
$a_0 = 0.0628$	$b_0 = 1$
$a_1 = 0.1256$	$b_1 = 1.2244$
$a_2 = 0.0628$	$b_2 = 0.483$

The transformed coefficients of 2-D decimating filter function obtained following the procedure previously described in Section 4.2 are given in Table 5.5 and Table 5.6, respectively for the filter in the z_2 dimension and for the filter in the z_1 dimension.

The resulting 2-D SC decimation filter network shown in Fig. 5.10(a) is formed by cascading two filters. Fig. 5.10(b) shows the switching waveforms. The IIR decimation filter in the z_2 dimension, with $M_2 = 3$, is followed by the IIR decimation filter in the z_1 dimension, with $M_1 = 2$. Thus, the required size of the DL memory blocks is only 1 / 3 of the size that would otherwise be required if non-multirate filter techniques were employed instead.

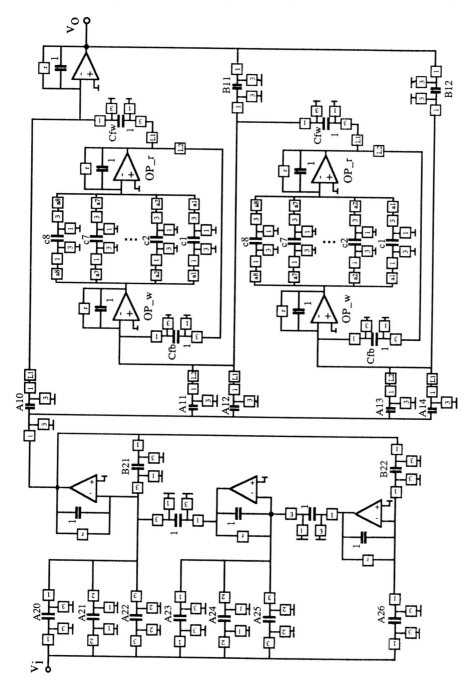

(a)

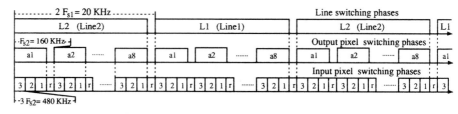

(b)

Fig. 5.10. (a) Schematic of the 2-D SC decimator with IIR filtering. (b) Switching waveforms.

Table 5.5
Coefficients of the modified 2-D decimating filter function in the z_2 dimension

Numerator Coefficients		Denominator Coefficients
$A_{20} = 0.0628$	$A_{24} = 0.1528$	$B_{20} = 1$
$A_{21} = 0.2025$	$A_{25} = 0.0665$	$B_{21} = -0.0614$
$A_{22} = 0.2805$	$A_{26} = 0.0147$	$B_{22} = 0.1127$
$A_{23} = 0.2417$		

Table 5.6
Coefficients of the modified 2-D decimating function in the z_1 dimension

Numerator Coefficients		Denominator Coefficients
$A_{10} = 0.0628$	$A_{13} = 0.1376$	$B_{10} = 1$
$A_{11} = 0.2025$	$A_{14} = 0.3034$	$B_{11} = -0.5332$
$A_{12} = 0.247$		$B_{12} = 0.2333$

The resulting computer simulated amplitude response is indicated in Fig. 5.11(a), for impulse sampling in both dimensions, and in Fig. 5.11(b) considering a S/H format for the output signal in the z_2 dimension.

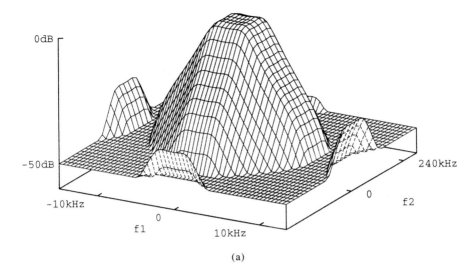

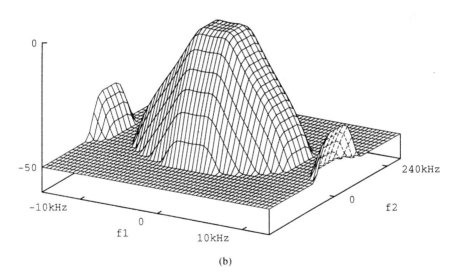

Fig. 5.11. Computer simulated amplitude responses of the 2-D SC decimation filter with IIR filtering function. (a) Impulse sampling. (b) Sampling-and-hold effect in the horizontal dimension.

5.6   SUMMARY

This chapter described various SC architectures for the implementation of 2-D decimation filters represented in polyphase-coefficient form, both with finite and infinite impulse responses. After reviewing the basic SC building blocks that realize the operations of coefficient multiplication and addition, we looked at the development of such SC architectures for different filter lengths and decimation factors in both dimensions. The complete design cycle, from specifications to the determination of nominal capacitance ratios, was illustrated giving the examples of an FIR 2-D filter and also of an IIR 2-D filter with separable transfer function. Computer simulations were presented to demonstrate the correct operation of such SC architectures. In the next chapter we shall consider in detail the practical integrated circuit implementation in CMOS technology of one such SC architecture.

REFERENCES

[5.1] P.E. Allen and D.R. Holberg, *"CMOS Analog Circuit Design"*, Holt, Rinehart and Winston, Inc., 1987.

[5.2] R. Gregorian and G. Temes, *"Analog Integrated Circuits for Signal Processing"*, John Wiley & Sons, Inc., 1986.

[5.3] Y. Tsividis and P. Antognetti, *"Design of MOS VLSI Circuits for Telecommunications"*, Prentice-Hall, Inc., Englewood Cliffs, NJ, 1985.

[5.4] R. Gregorian and W.E. Nicholson, "SC Decimation and Interpolation Circuits", *IEEE Transactions on Circuits Systems*, vol. 27, June 1980.

[5.5] D.C. Grunigen, U.W. Brugger and G.S. Moschytz, "A Simple SC Decimation Circuit", *Electronics Letters*, vol. 17, January 1981.

[5.6] D.C. Grunigen, R. Sigg, M. Ludwig, U.W. Brugger, G.S. Moschytz and H. Melchior, "Integrated SC Low-Pass Filter with Combined Anti-Aliasing Decimation Filter for Low Frequencies", *IEEE J. of Solid-State Circuits*, Vol. 17, pp. 1024-1028, December 1982.

[5.7] J. E. Franca, "Non Recursive Polyphase Switched-Capacitor Decimators and Interpolators", *IEEE Transactions on Circuits and Systems*, vol. CAS-32, no. 9, pp. 877-887, September 1985.

[5.8] R.P. Martins, J.E. Franca and F. Maloberti, "An Optimum CMOS SC Anti-Aliasing Decimating Filter", *IEEE J. of Solid-State Circuits*, Vol. 28, No. 9, pp. 962-970, September 1993.

[5.9] Paulo J. Santos, J. E. Franca and J.A. Martins, "SC Decimators Combining Low-Sensitivity Ladder Structures with High-Speed Polyphase Networks", *IEEE Transactions on Circuits Systems - II: Analog and Digital Processing*, vol. 43, no. 1, pp. 31-38, January 1996.

[5.10] J. E. Franca and S. Santos, "FIR Switched-Capacitor Decimators with Active-Delayed Block Polyphase Structures", *IEEE Transactions on Circuits Systems*, vol. 35, no. 8, pp. 1033-1037, August 1988.

[5.11] SWITCAP-2 User's Manual, Columbia University.

6

A REAL-TIME 2-D ANALOG MULTIRATE IMAGE PROCESSOR IN 1.0-μm CMOS TECHNOLOGY

6.1 INTRODUCTION

This chapter demonstrates the practical integrated circuit implementation of a multirate image processor employing the SC architectures described in the previous chapter. The filter specifications and corresponding multirate-transformed transfer functions are given in Section 6.2. Section 6.3 discusses the design at the architectural level of the SC structure employed for the horizontal decimating filter. Section 6.4 presents the architecture level design of the SC vertical filter and associated DL memory blocks. Then, in Section 6.5, we examine the design of the various building blocks needed for integrated circuit implementation, namely the fully-differential OTA as well as the analog storage cells and the analog multiplexers and demultiplexers for memory addressing. Section 6.6 gives the results of the experimental characterization of the prototype filter chip. Finally, Section 6.7 draws the conclusions of the chapter.

6.2 2-D MULTIRATE FILTER DESIGN

The basic architecture of the 2-D processor, shown in Fig. 6.1(a), comprises a 2nd-order Butterworth 2-fold lowpass decimation filter in the horizontal dimension and a 2nd-order Butterworth lowpass filter in the vertical dimension. The horizontal decimating filter processes the pixels of each line whereas the vertical filter processes the lines of each frame.

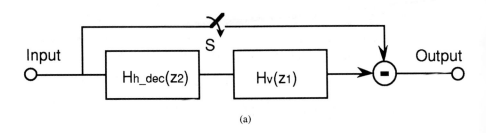

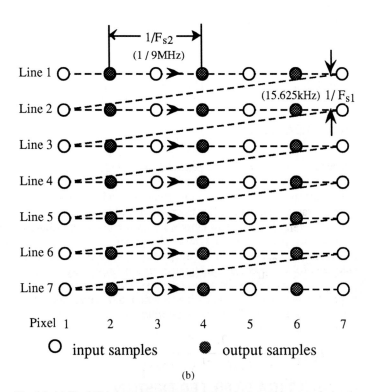

Fig. 6.1. (a) The 2-D image processor comprises an horizontal lowpass decimating filter and a vertical lowpass filter. (b) The line memory depth of the vertical filter is halved due to the horizontal 2-fold decimating filtering function.

As schematically illustrated in Fig. 6.1(b), the reduction of the horizontal pixel sampling rate from $2F_{s2} = 18$ MHz at the input to $F_{s2} = 9$ MHz at the output allows a 2-fold reduction of the line memory depth that is needed for the vertical line filtering function at $F_{s1} = 15.625$ kHz. When switch S is open the two lowpass filtering functions in cascade produce an overall (2 × 2)nd-order lowpass filtering function. By closing switch S the complementary (2 × 2)nd-order highpass filtering function results from the subtraction of the lowpass filtered signal from the incoming input signal.

The horizontal decimating filter in the above 2-D image processor is designed to meet the specifications given in Table 6.1.

Table 6.1
Specifications of the horizontal decimating filter

Filter order	2
Passband frequency	1.7 MHz @ 3 dB
Stopband frequency	4.5 MHz @ 20 dB
Sampling rate	18 MHz
Decimated sampling rate	9 MHz
Horizontal decimation ratio	2

Considering the bilinear z-transform and after applying the multirate transformation [6.1] for the given sampling rate reduction $M_2 = 2$ we arrive at the expression

$$(6.1) \quad H_{h\_dec}(z_2) = \frac{0.061 + 0.196 \, z_2^{-1} + 0.234 \, z_2^{-2} + 0.126 \, z_2^{-3} + 0.027 \, z_2^{-4}}{1 - 0.543 z_2^{-2} + 0.187 \, z_2^{-4}},$$

describing the z_2-transfer function of the horizontal decimating filter and where the unit-delay term z_2^{-1} refers to the input sampling frequency of the horizontal pixels.

The vertical filter, in turn, is designed to meet the specifications given in Table 6.2. The resulting z_1-transfer function obtained based on the impulse-invariant z-transform is expressed as

$$(6.2) \quad H_v(z_1) = \frac{0.347 \, z_1^{-1}}{1 - 0.953 \, z_1^{-1} + 0.320 \, z_1^{-2}},$$

where the unit delay term z_1^{-1} refers to the line sampling frequency.

Table 6.2
Specifications of the vertical filter

Filter order	2
Passband frequency	2 kHz @ 3 dB
Stopband frequency	6.3 kHz @ 20 dB
Vertical sampling rate	15.625 kHz
Operating speed	9 MHz
DL storage cells	570 cells

From the above expressions (6.1) and (6.2) the overall 2-D decimation filter function can be expressed as

$$(6.3) \quad H_{2D\_dec}(z_1, z_2) = H_{h\_dec}(z_2) H_v(z_1),$$

for the lowpass output, and as

$$(6.4) \quad H_{multi\_HP}(z_1, z_2) = 1 - H_{h\_dec}(z_2) \cdot H_v(z_1),$$

for the highpass output. Next, we shall consider the design of the SC circuits that implement the above transfer functions for both the horizontal and vertical filters.

6.3 SC HORIZONTAL DECIMATING FILTER

The SC horizontal high frequency lowpass filter is based on a low-sensitivity biquadratic filter structure [6.2 – 6.3]. To minimize noise and various parasitic effects that are particularly visible at high frequency, the fully-differential circuit architecture shown in Fig. 6.2 is adopted [6.4 – 6.7]. The circuit uses only two clock phases, ϕ_1 and ϕ_2, together with their complementary phases, ϕ_{n1} and ϕ_{n2}. Its transfer function can be expressed by [6.3]

$$(6.5) \quad H_{h\_dec}(z_2) = \frac{(C'_{21} + C''_{21})z_2^2 + (C_{21} C_{23} - C'_{21} - 2C''_{21})z_2 + C''_{21}}{(1 + C_{24})z_2^2 + (C_{22} C_{23} - C_{24} - 2)z_2 + 1},$$

where coefficients C_{2j} represent the normalized capacitance values in the biquads, i.e. the ratios between the input capacitance and the integrating capacitance of the amplifiers. By writing the previous z_2-transfer function (6.1) in terms of polyphase-coefficient functions

(6.6) ... $H_{h\_dec}(z_2) = \dfrac{(A_0 + A_1 z_2^{-1}) + (A_2 + A_3 z_2^{-1}) z_2^{-2} + A_4 (z_2^{-2})^2}{1 + B_1(z_2^{-2}) + B_2(z_2^{-2})^2}$,

and equating it to the above transfer function (6.5) of the biquad we arrive at the following design equations

(6.7-a) ... $C''_{21} = \dfrac{A_4}{B_2}$,

(6.7-b) ... $C'_{21} = \left(\dfrac{A_0 - A_4}{B_2}\right) + \dfrac{A_1}{B_2} z_2^{-1}$,

(6.7-c) ... $C_{22} = \dfrac{1 + B_1 + B_2}{B_2}$

(6.7-d) ... $C'_{21} = \left(\dfrac{A_0 + A_2 + A_4}{B_2}\right) + \dfrac{A_1 + A_3}{B_2} z_2^{-1}$, and

(6.7-e) ... $C_{23} = 1$, $C_A = 1$, $C_B = 1$.

In the expressions (6.7-b) and (6.7-d) above we can notice that each of C_{21} and C''_{21} are composed of two terms, one proportional to $z_2^{\,0}$ and the other proportional to z_2^{-1}, corresponding to the polyphase-coefficient representation in (6.6).

After scaling the internal voltages for maximum signal handling, scaling capacitance for minimizing the capacitance spread-ratio and checking the capacitance loading effect of the OTA with finite DC-gain, we obtain the denormalized capacitance values (in pF) also indicated in Fig. 6.2.

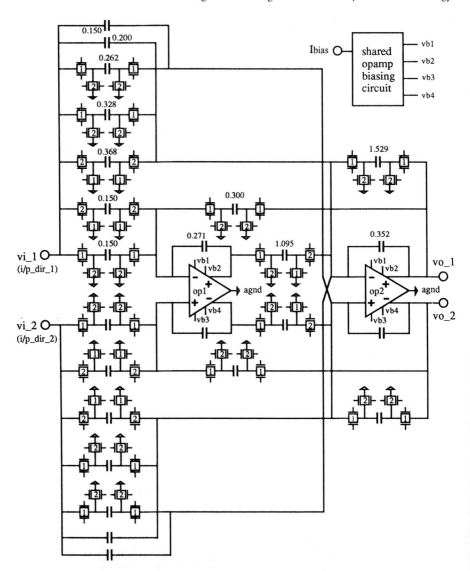

Fig. 6.2. Circuit diagram and capacitance values (in pF) of the fully-differential SC horizontal decimation filter.

6.4 SC VERTICAL FILTER AND ASSOCIATED DELAY-LINE MEMORY BLOCKS

6.4.1 Delay-Line Memory Blocks

In order to store the full line information produced at the output of the horizontal filter we consider the delay-line memory block schematically illustrated in Fig. 6.3(a), for a simplified single-ended version.

This comprises two OTAs and a storage capacitor-array and is operated by the same two-phase non-overlapping clock phases ϕ_1 and ϕ_2 previously used in the horizontal filter. The selection of the active Read (R) and Write (W) modes is performed by switches *Write* and *Read*.

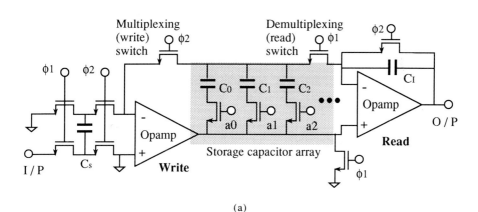

(a)

(b)

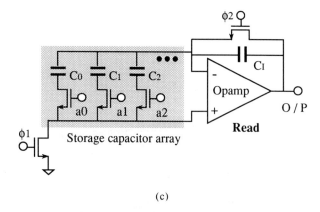

(c)

Fig. 6.3. (a) Simplified (single-ended) circuit diagram of the DL memory block and corresponding (b) Write and (c) Read modes.

In the Write mode, the circuit is configured as shown in Fig. 6.3(b). During the clock phase ϕ_1, an input sample is taken and stored on C_s. During clock phase ϕ_2 the charge is transferred from C_s to a selected storage capacitor by configuring the storage capacitor as a feedback element around one OTA (OP_write) [6.8 – 6.9]. Selection of the storage capacitor is accomplished by closing the appropriate MOS switches clocked by phases a_0, a_1, a_2 and so forth. In the Read mode, the circuit is reconfigured as shown in Fig. 3(c). During clock phase ϕ_1, the sample from the selected cell is transferred to the integrating capacitor C_I by the other OTA (OP_read). The next clock phase ϕ_2 is used to reset the integrating capacitor C_I before reading a new position from the storage capacitor-array.

Because in the above architecture the cell select transistor is employed as the sampling switch during the write operation, careful placement of this switch ensures that signal-dependent charge injection is minimized. Cell-to-cell matching of the storage capacitors does not affect, to first order, the operation of the circuit [6.9]. The timing of the circuit is such that for a given storage cell, the read operation is performed immediately prior to writing a new sample. This has the advantage of resetting the storage capacitor so that a separate reset phase is not necessary immediately prior to the write phase.

The actual integrated circuit implementation of the DL memory block is highly constrained by the associated large parasitic capacitances. For a simple linear-type structure of the capacitor arrays, the total source-drain diffusion parasitic capacitance of the selected switches as well as the total bottom plate capacitance of the unselected storage capacitors can easily be of the order of 10 – 15 pF and thus placing extremely stringent demands on the power capability of the OTAs to achieve the required fast access time.

Multirate Switched-Capacitor Circuits for 2-D Signal Processing 107

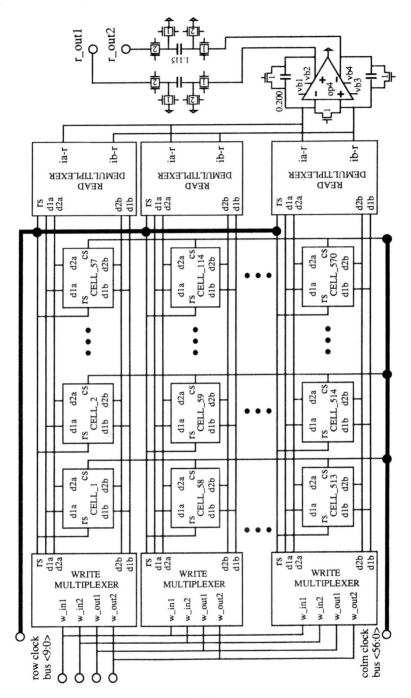

(a)

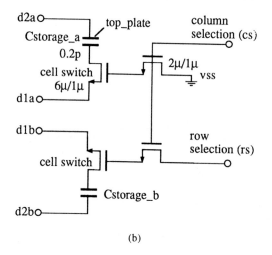

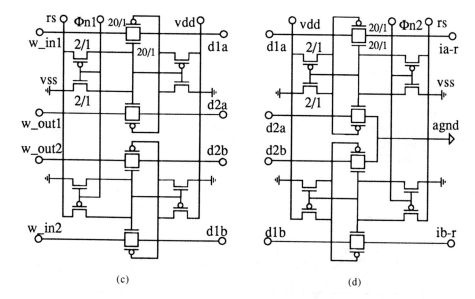

Fig. 6.4. Fully-differential DL memory block. (a) Architecture and (b) storage cell. (c) Multiplexer circuit for writing. (d) Demultiplexer circuit for reading.

By adopting instead a matrix-type structure of the capacitor arrays [6.8 – 6.9] with 10 rows and 57 columns, such total parasitic capacitance can be reduced to as low as 1 – 1.5 pF. Besides the basic analog storage cells, this solution also requires 2-dimensional (row x column) multiplexers and demultiplexers for cell access, as shown in the schematic diagram of Fig. 6.4(a) representing the overall organization of the DL memory blocks. The smaller square in the figure represents a fully-differential storage cell, whose schematic is drawn as in Fig. 6.4(b), whereas the multiplexer and demultiplexer schematics are given, respectively, in Fig. 6.4(c) and Fig. 6.4(d).

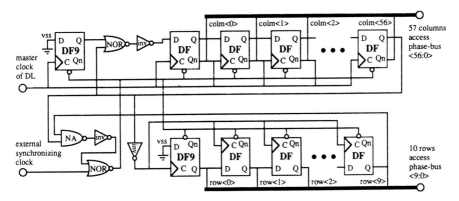

Fig. 6.5. DL address generation circuit.

The selection of a particular storage cell is accomplished by the two circular address generators schematically illustrated in Fig. 6.5. One output clock bus <0:9> generates circularly 10 successive periods to scan 10 rows of the full DL memory block, while the other output clock bus <0:56> produces 57 successive time slots to scan 57 columns of the DL memory block. The overall clock circuit is triggered by an external master clock. An external DL synchronizing port, *Sync*, is included to synchronize the filter operation with a variety of image sampling rates.

6.4.2 SC Vertical Filter

The SC implementation of the vertical filter employs a biquadratic structure, also fully differential, adequately modified to include the DL memory blocks described above. The resulting circuit diagram is shown in Fig. 6.6, where the two rectangles together with the associated OTAs for Write and Read modes represent the two previously described DL memory blocks. In order to allow independent testing of a DL memory block alone the circuit includes an additional clock phase ϕ_t with the same timing as ϕ_2, but which uses a different connectivity in the circuit to isolate the DL_Memory_2 when tested off-chip. The resulting transfer function of such biquadratic filter structure can be expressed by

$$(6.8) \ldots H_v(z_1) = \frac{C_{11} C_{13} z_1^{-1}}{(1 + C_{14}) z_1^2 + (C_{12} C_{13} - C_{14} - 2) z_1 + 1},$$

where the coefficients C_{1j} represent the capacitors of the biquad normalized to the integrating capacitors of the OTAs. By equating the above transfer function to the vertical filter transfer function (6.2), and then carrying out the appropriate scaling operations for maximum signal handling, minimum spread of the capacitance ratios, we obtain the capacitance values (in pF) also indicated in Fig. 6.6.

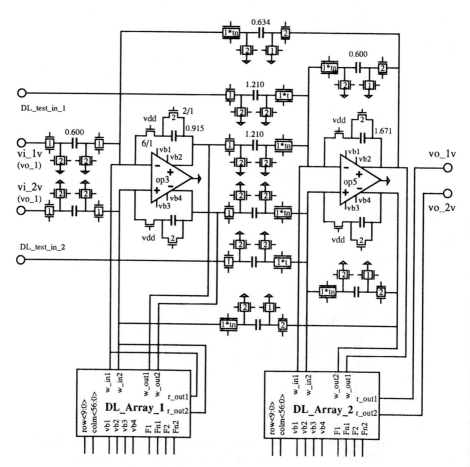

Fig. 6.6. Circuit diagram and capacitance values (in pF) of the fully-differential SC vertical filter.

6.5 INTEGRATED CIRCUIT IMPLEMENTATION

6.5.1 Evaluation of Key Parasitic Effects and Gain-Speed Limitations

A closer examination of the DL memory block circuit during the write phase shows that the non-zero resistances of the MOS switches may adversely affect the stability and settling of the OTA. To illustrate this, Fig. 6.7 shows a simplified, equivalent circuit in the Write mode with the relevant resistances due to the select switch in the cell and the row multiplexers along with the parasitic capacitances. Assuming a single pole model for the OTA, the complete feedback system has four poles and hence the analysis of its settling becomes rather difficult [6.8 - 6.9]. In particular, the poles formed by the resistance R_{mux} of the multiplexer switches and array capacitance $C_{stray-mux}$ can be especially troublesome if they fall inside the bandwidth of interest. However, in contrast with the switch in the storage cell, the switches in the multiplexer can be made reasonably large without adversely impacting the overall circuit area. Thus, to simplify the design, the multiplexer switches are made large enough such that the pole due to the switch resistances Rmux falls well beyond the remaining two poles, one of which is due to the output loading of the OTA.

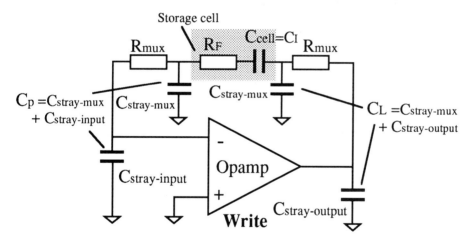

Fig. 6.7. Simplified DL memory block circuit in the Write mode with parasitic capacitances and resistances.

The second pole in the equivalent circuit in Fig. 6.7 results from the combination of the feedback capacitance, which is the storage cell capacitor C_{cell} that is equal to the integrating capacitor C_I of Read OTA , together with the storage cell switch resistance R_F. Ignoring the poles due to the multiplexer switch resistance

R_{mux}, the resulting settling behavior of the system can be described by a 2nd-order rational function $N(s)/D(s)$ where the denominator $D(s)$ can be expressed as

$$(6.9) \quad D(s) = 1 + \frac{s}{g_m}\left[C_p + C_L + \left(\frac{C_p C_L}{C_I}\right)\right] + s^2\left(\frac{C_p C_L R_F}{g_m}\right),$$

yielding, in turn, the system time-constant

$$(6.10\text{-a}) \quad \tau = \frac{2Q}{\omega_0\left(1 - \sqrt{1 - 4Q^2}\right)},$$

as well as the pole quality-factor

$$(6.10\text{-b}) \quad Q = \frac{\sqrt{C_L C_p R_F g_m}}{C_L + C_p + \frac{C_L C_p}{C_I}},$$

and pole frequency

$$(6.10\text{-c}) \quad \omega_0 = \sqrt{\frac{g_m}{C_L C_p R_F}}.$$

Based on the stability analysis of the above 2nd-order system, with both C_P and C_L between 1 pF and 1.5 pF, and R_F = 1 kΩ, a value of Q = 0.25 can be optimally defined to achieve an OTA with g_m = 1 mS and C_I = 0.2 pF. This is within the reach of single-stage folded-cascode OTA architectures and, for comparable speed requirements, it is much smaller than the 100 mS transconductance required in [6.9]. By making the DL storage cell C_I large relative to C_P will affect the size and power requirements of the circuit, whereas a very small value of C_I is limited by process matching accuracy.

6.5.2 OTAs and Biasing Circuit

For the target decimated settling time of 110 ns, 8-bit settling accuracy and total capacitive load of 2.5 pF, the OTA itself must exhibit worst-case minimum 62 dB DC gain, 13 ns settling time and 120 MHz unity gain bandwidth. This can be achieved using fully-differential folded-cascode structures with SC common-mode feedback [6.5], [6.10 – 6.11]. The resulting circuit schematic diagram is shown in Fig. 6.8(a) whereas in Fig. 6.8(b) we can observe the master biasing circuit that is employed for current biasing of all the OTAs in the system.

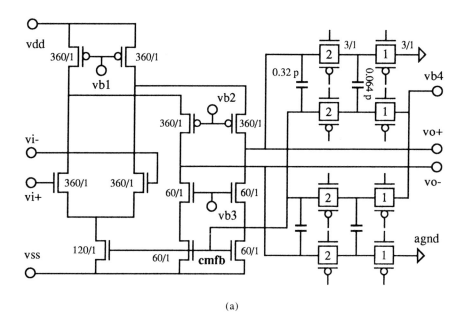

(a)

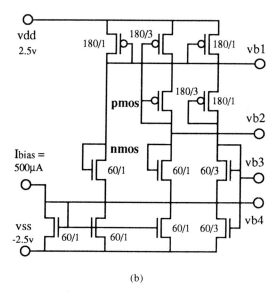

(b)

Fig. 6.8. (a) Fully-differential folded-cascode OTA with (b) master current biasing circuit.

114   6. A Real-Time 2-D Analog Multirate Image Processor in 1.0-μm CMOS Technology

6.5.3 Overall System Architecture and Chip Floor Plan

By combining together the filter structures and key building blocks described before we obtain the block diagram illustrated in Fig. 6.9(a) for the complete 2-D image processor. The corresponding chip floor plan, depicted in Fig. 6.9(b), is established as symmetrical as possible to increase the immunity against unavoidable process and accuracy limitations [6.14]. The two large blocks in the lower half portion of the system are the DL memory blocks, each containing 570 fully-differential storage cells. This gives a total of 4 x 570 storage capacitors and approximately 8 x 570 switches that are needed for the implementation of the DL memory blocks alone. The row multiplexers and demultiplexers, respectively for writing and reading the DL memory blocks, are placed at the top of the analog storage area. The column address generators are placed between the two DL memory blocks, whereas two equal row address generators are placed below each DL memory block. The 7 OTAs employed in the whole system are located in the upper half portion of the chip floor plan. From left to right, the first 2 OTAs serve the horizontal (z_2) decimating filter, whereas the next 4 OTAs are employed in the vertical (z_1) filter and associated DL memory blocks. The last OTA is associated with the highpass filtering response of the system and which, as seen in Fig. 6.1(a), is produced by subtracting the lowpass filtered signal from the direct input signal.

For facilitating the full testing of the system, several connectivity options are provided to the various blocks. For example, the horizontal decimation filter output can be connected externally to the input of the vertical filter input where, in turn, the DL memory blocks can also be configured for stand-alone testing. The summing operation around OP7 can also be configured off-chip.

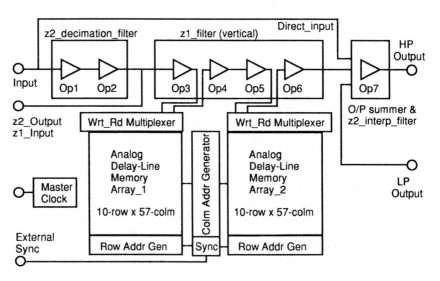

(a)

Multirate Switched-Capacitor Circuits for 2-D Signal Processing   115

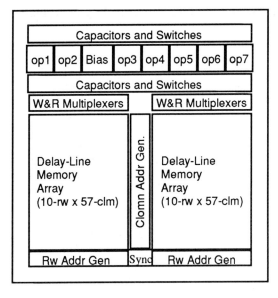

(b)

Fig. 6.9. (a) Block diagram and (b) chip floor plan of the complete 2-D multirate image processor.

The master clocks ϕ_1 and ϕ_2 for the overall filter system are also generated externally to allow variability of the master clocks duration to the column address clocks that require about 8 ns delay from the master. A synchronization signal pad is also externally available to apply different line delay times.

6.6 EXPERIMENTAL CHARACTERIZATION

The experimental 2-D multirate processor describe above has been integrated in a 1.0-μm CMOS technology with 2 layers of metal and double-poly layers. Fig. 6.10 shows the microphotograph of the resulting prototype chip occupying a total area (active core) of only 2.5 x 3.0 mm^2.

The testing set-up used for experimental characterization of the prototype chip is illustrated in Fig. 6.11. Besides the auxiliary circuits for single-ended(differential)-to-differential(single-ended) conversion, and the usual equipment for frequency response characterization, it also includes a PAL video camera and a video monitor to test the real-time image processing capability of the processor.

116   6. A Real-Time 2-D Analog Multirate Image Processor in 1.0-μm CMOS Technology

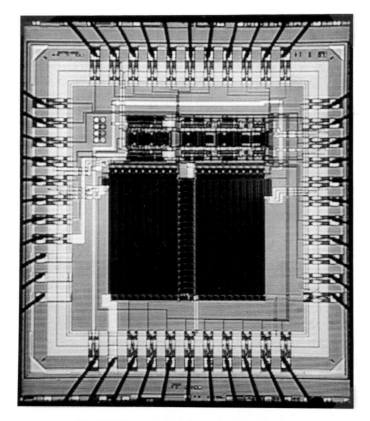

Fig. 6.10. Microphotograph of the prototype 2-D filter.

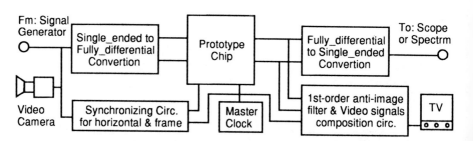

Fig. 6.11. Experimental set-up for chip testing and characterization.

The measured amplitude responses of the horizontal filter alone are plotted in Fig. 6.12 for sampling frequencies F_{s2} from 10 MHz to 18 MHz. Because of the $M_2 = 2$ decimation factor, the output baseband extends from DC to $f/F_{s2} = 0.25$. For comparison purposes, the electrical simulated response of the extracted layout of the circuit is also indicated. Note, however, that such simulation does not include the sample and hold effect that is present at the output of an SC filter and which is actually seen in the measured responses. The measured and simulated responses agree well in the filter passband up to the -3 dB cut-off frequency at $f/F_{s2} = 0.094$ (from Table 6.1, this is 1.7 MHz at 18 MHz sampling). Specifically, the -3 dB cut-off frequency error is -0.2% when the filter is running at the NTSC compatible sampling frequency of 14.314 MHz and -0.35% at the PAL compatible sampling frequency of 17.718 MHz. Just before the Nyquist point of $f/F_{s2} = 0.25$ (from Table 6.1, this is 4.5 MHz at 18 MHz) the amplitude response deviation increases by about 2 dB the minimum stopband attenuation of the filter running at 14.314 MHz, while at 17.718 MHz sampling this excess stopband attenuation increases to about 5 dB.

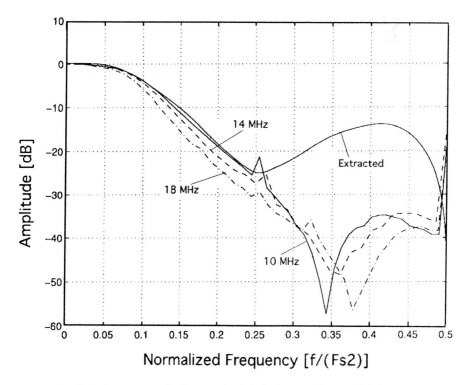

Fig. 6.12. Frequency amplitude responses of the implemented horizontal decimation filter.

118   6. A Real-Time 2-D Analog Multirate Image Processor in 1.0-μm CMOS Technology

(a)

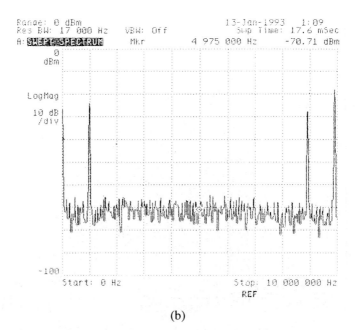

(b)

Fig. 6.13. Measured (a) amplitude response and (b) single-frequency response of the DL memory block.

Multirate Switched-Capacitor Circuits for 2-D Signal Processing

For the DL memory block alone clocked at 20 MHz, we obtain the measured amplitude response shown in Fig. 13(a). This is almost ideally flat (apart from the 20 dB loss due to signal-level restrictions of the equipment) in the video signal bandwidth from up to about 5 MHz. Fig. 13(b) shows the measured response to a single-frequency input test signal. The overall measured amplitude response of the 2-D processor in the lowpass filtering mode, given in Fig. 14, agrees well with the expected nominal responses considering the additional effect of the S/H signals. The main experimental characteristics obtained for the prototype chip are summarized in Table 6.3.

Table 6.3
Summary of measured performance characteristics of the CMOS prototype chip

Maximum Horizontal Filter Clock Rate	40 MHz
2-D Decimation Filter Clock	18 MHz
Power Consumption	85 mw @ 18 MHz
Active Area	2.5 x 3.0 mm^2
Dynamic Range (signal to random noise)	> 50 dB
Signal Swing	> 2 V$_{p-p}$

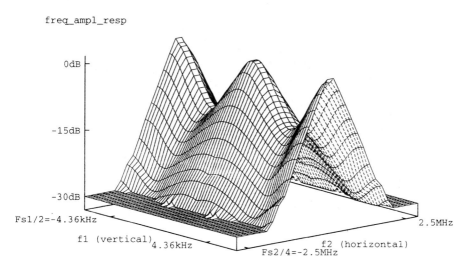

Fig. 6.14. Experimental frequency responses of the 2-D decimation filter.

6. A Real-Time 2-D Analog Multirate Image Processor in 1.0-μm CMOS Technology

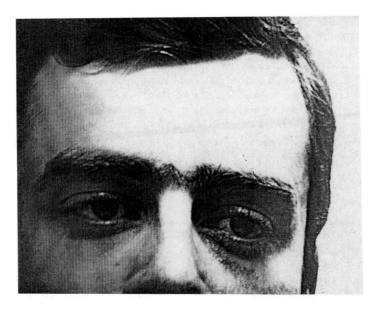

(a)

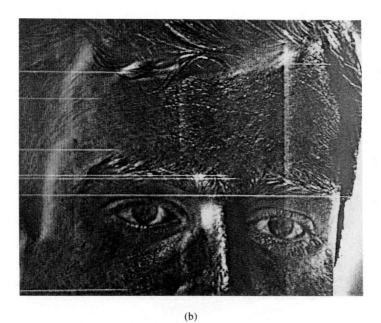

(b)

Fig. 6.15. Paulo's images. (a) Original PAL image and (b) its highpass filtered version from the prototype chip.

Multirate Switched-Capacitor Circuits for 2-D Signal Processing 121

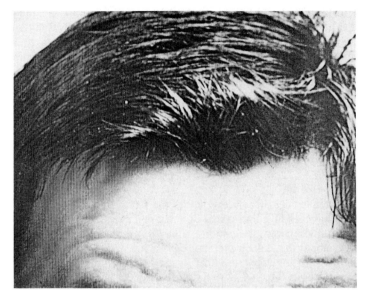

(a)

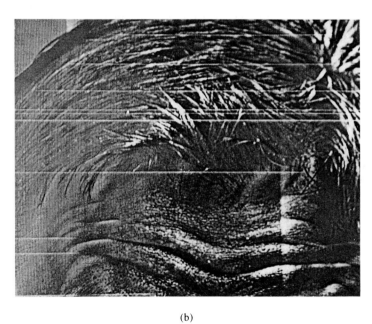

(b)

Fig. 6.16. The forehead images. (a) Input PAL image and (b) output highpass version from the prototype chip.

After evaluating the main constituting building blocks, the chip is configured as an (2 x 2)nd-order highpass filter whose input signals is taken from the PAL camera and whose output is displayed in the TV monitor. This requires some auxiliary circuits for decomposing and reconstructing the video synchronization signals, as indicated in Fig. 11, as well as a simple 1st-order anti-image filter to smooth the output of the 2-D filter. Difficulty in achieving perfect synchronization with the line interlacing signals of the commercial TV monitor resulted in some spurious lines appearing in the projected image signal of the 2-D processor. Two examples of photographs of the input and output image signals displayed in the TV monitor are shown, respectively, in Fig. 15 and Fig. 16, respectively. The Fig. 15(b) and Fig. 16(b) show the expected edge enhancement effects due to the highpass filtering function.

For completeness, Table 6.4 shows the relative cost of implementation of this fully analog multirate chip for real-time image processing, from the viewpoints of silicon area and power dissipation, vis-a-vis the cost of implementing in a technology of about the same generation as the one employed herein the analog-digital and digital-analog converters that would be needed to support an alternative fully digital processing core [6.13 – 6.14]. This clearly indicates that for moderate accuracy and low to moderate complexity of the processing function, a fully multirate analog implementation has a potential to achieve a more competitive implementation than an alternative digital VLSI implementation. However, for high accuracy and/or higher processing complexity, not only the relative overhead cost of the front-end and back-end converters will diminish but also the implementation of the processing core in digital VLSI will benefit more of technology scaling to achieve higher density of integration.

Table 6.4
Cost of implementation of the fully analog real-time image processing chip and the conversion blocks that would be needed to support a fully digital processing core

Function	Equivalent Throughput	Silicon Area	Power Dissipation	Technology
Analog image processor	Nominal: 144 Mbit/sec Maximum: 320 Mbit/sec	7.5 mm^2	85 mW at 5 V and 18 MHz	1.0-µm CMOS
ADC [6.15]	200 Mbit/sec	10.56 mm^2	35 mW at 3 V and 20 MHz	1.2-µm CMOS
DAC [6.16]	400 Mbit/sec	0.4 mm^2	14 mW at 5 V and 40 MHz	0.8-µm CMOS

6.7 SUMMARY

A real-time 2-D switched-capacitor image processor prototype chip that realizes (2 × 2)nd order recursive lowpass and highpass filtering functions was designed in a 1.0-μm CMOS double-poly and double-metal technology employing an efficient horizontal decimating filter to reduce the size of the DL memory blocks. Careful analysis of parasitic and speed limitation effects associated with the write and read operation modes of the storage cells allowed to design only one type of OTA, with nominal 1 mS DC transconductance and over 100 MHz unity gain bandwidth, for both the vertical filter and associated DL memory blocks and the horizontal filter. Because of the 2-fold horizontal decimation the size of the delay-line memory blocks is reduced to only 570 storage cells, compared to 1140 cells that would be needed in a non-multirate system. The active core area is a mere 2.5 × 3.0 mm^2 and at 5 V supply and nominal 18 MHz input sampling the chip dissipates only 85 mW.

REFERENCES

[6.1] R. Crochiere and L.R. Rabiner, *"Multirate Digital Signal Processing"*, Prentice-Hall, Englewood Cliffs, NJ, 1983.

[6.2] P.E. Allen and D.R. Holberg, *"CMOS Analog Circuit Design"*, Holt, Rinehart and Winston, Inc., 1987.

[6.3] R. Gregorian and G. Temes, *"Analog Integrated Circuits for Signal Processing"*, John Wiley & Sons, Inc., 1986.

[6.4] Y. Tsividis and P. Antognetti, *"Design of MOS VLSI Circuits for Telecommunications"*, Prentice-Hall, Inc., Englewood Cliffs, NJ, 1985.

[6.5] T.C. Choi, R.T. Kaneshiro, R.W. Brodersen, P. Gray, W.B. Jett and M. Wilcox, "High-Frequency CMOS SC Filters for Communications Application", *IEEE Journal of Solid-State Circuits*, vol. 18, no.6, pp. 652-664, December 1983.

[6.6] M.S. Tawfik and P. Senn, "A 3.6-MHz Cutoff Frequency CMOS Elliptic Lowpass SC Ladder Filter for Video Communication", *IEEE Journal of Solid-State Circuits*, vol. 22, no.3, pp. 378-384, June 1987.

[6.7] B.S. Song, "A 10.7-MHz SC Bandpass Filter", *IEEE Journal of Solid-State Circuits*, vol. 24, no.2, pp. 320-324, April 1989.

[6.8] K.Matsui, T.Matsuura, S.Fukasawa, Y.Izawa, Y.Toba, N.Miyake and K.Nagasawa, "CMOS Video Filters Using SC 14-MHz Circuits", *IEEE Journal of Solid-State Circuits*, Vol. sc-20, No. 6, pp. 1096-1101, December 1985.

[6.9] K. A. Nishimura and P. R. Gray, "A Monolithic Analog Video Comb Filter in 1.2-μm CMOS", *IEEE J. of Solid-State Circuits*, Vol. 28, No. 12, pp. 1331-1339, December 1993.

[6.10] G.T. Uehara and P.R. Gray, "Practical Aspects of High-Speed SC Decimation Filter Implementation", *Proc. IEEE International Symposium on Circuits and Systems*, pp. 2308-2311, May 1992.

[6.11] M. Steyaert and W. Sansen, "Opamp Design towards Maximum Gain-Bandwidth", *Workshop on Advances in Analog Circuit Design*, pp. 59-80, April 1992.

[6.12] F. Maloberti, "Layout for Analog and Mixed A/D Systems", *GCSI Group Lectures*, June 1992.

[6.13] T.B. Cho and P.R. Gray, "A 10-b, 20-MS/s, 35 mW Pipeline A/D Converter", *IEEE Journal Solid-State Circuits*, Vol. 30, No. 3, pp. March 1995.

[6.14] C.A. Bastiaansen, D.W. Groeneveld and H. Schouwennaars, "A 10-b, 40-MS/s, 0.8-μm CMOS Current-Output D/A Converter", *IEEE Journal Solid-State Circuits*, Vol. 26, No. 7, pp. 917-921, July 1991.